CLEAR TECHNICAL REPORTS

CLEAR TECHNICAL REPORTS

WILLIAM A. DAMERST

The Pennsylvania State University

HARCOURT BRACE JOVANOVICH, INC. [HBJ]

New York   Chicago   San Francisco   Atlanta

ISBN: 0-15-507691-4

Library of Congress Catalog Card Number: 72-77041

Printed in the United States of America

For Dibby and the rest of our family

CONTENTS

PREFACE   xi

Part 1

UNDERSTANDING CLEAR COMMUNICATION   1

Chapter One
THE BASIC PROCESS AND READERS' WANTS   3

Understanding the Communication Process—The Key to Good
   Practical Writing   4
"Old" Technical Terms and a New One—And Their
   Importance   5
Anticipating Obstacles in the Communication Process   6
Obstacles to Satisfying the Readers' Wants   7
   Exercises   8

Chapter Two
OBSTACLES TO UNDERSTANDING THE MESSAGE   11

Inaccuracy   11
Unclear Reason for Writing   13
Incompleteness   15
Overwriting   16
Lack of Candor   17
Lack of Empathy   18
Omission of Summary   20
   Exercises   21

Chapter Three
OBSTACLES TO ACCEPTING THE FORMS 26

Overall Forms 26
Orders of Development 29
Outlines As Headings 31
Paragraphs 34
   Exercises 38

Chapter Four
OBSTACLES TO UNDERSTANDING THE LANGUAGE 42

Sentences, Clauses, and Phrases 43
Types of Sentences 44
Other Ways of Classifying Sentences 49
Phrases 53
Awkwardness 54
The Words of Language 55
Readability Formulas 59
   Exercises 64

Part 2
MASTERING THE SKILLS OF TECHNICAL WRITING 67

Chapter Five
RESEARCH AND INTERPRETATION 69

The Meanings of Interpretation 70
Research 71
Note Taking and Documentation 76
Logic and Illogic 83
Special Uses of Interpretation During Investigative Work 87
Importance of Standards in All Interpretations 91
Importance of Using Imagination in Interpretative Work 91
   Exercises 93

Chapter Six
DEFINITION AND DESCRIPTION 98

Definition 99
Classification and Partition 107

Description   109
    Exercises   118

Chapter Seven
SUMMARIES, LISTS, AND GRAPHIC AIDS   121

Summaries   121
Lists   127
Graphic Aids   128
    Exercises   141

Part 3

PRESENTING COMMON FINISHED PRODUCTS 147

Chapter Eight
LETTERS   149

Layout   149
Requests and Inquiries, and Replies   157
Orders and Invitations to Bid   161
Claims and Complaints, and Replies   166
Notices, Announcements, and Other Information-Giving
    Letters   169
Tone and Letter Language   170
    Exercises   174

Chapter Nine
MEMOS AND INFORMAL REPORTS   180

Memos   180
Informal Reports   194
The Importance of Explicit Statements of Purpose and Scope   206
    Exercises   208

Chapter Ten
FORMAL REPORTS AND PROPOSALS   214

Reasons for Formal Presentations   214
The Formal Report   215
A Specimen Formal Report   220
The Formal Proposal   235
    Exercises   241

Part 4

PRESENTING SPECIAL FINISHED PRODUCTS 245

Chapter Eleven
ARTICLES 247

Recognizing the Benefits of Authorship   247
Finding a Subject   248
Identifying Reader Groups   249
Adapting Material for Readers   250
Writing Articles Regularly   257
Clearing Articles with Employers   258
    Exercises   258

Chapter Twelve
THE JOB APPLICATION 260

The Résumé   261
Notes about Résumé Style   265
Use of the Résumé Instead of a Company Form   266
The Covering Letter   266
Follow-ups   271
    Exercises   272

Part 5

A TECHNICAL WRITER'S HANDBOOK 277

Abbreviations (AB)   280
Numbers (NUM)   287
Spelling (SP)   288
Explanations of Grammatical Terms (GT)   293
Common Errors in Grammar, Diction, and Punctuation   303
(SU) Violations of Sentence Unity   303
(SC) Violations of Sentence Coherence   305
(D)   Errors in Diction—Word Choice and Usage   313
(P)   Errors in Punctuation   315
    Exercises   324

INDEX 329

PREFACE

The idea behind this book—that technical writing must first of all be *clear*—came to me when I first taught and consulted in industry. Working daily with engineers, chemists, physicists, and others, I found that the greatest weakness in on-the-job writing—not being clear to the reader—was evident in many of the messages that were written. Indeed, in many cases the Ph.D.'s in industry had the same difficulties in making their messages clear that the technicians had.

Now, more than twelve years later—having worked with thousands of students in industry, in the classroom, and through correspondence courses—I have given shape to the idea behind *Clear Technical Reports.* That shape enables the student to make use of many forms for communication; necessarily, the form varies as the situation varies. However, the student will learn that the basic *principles* of communication do not vary. Through the book, therefore, the student will be prepared to handle effectively *any* writing problem he may have in his career work.

The book is divided into five parts. Part 1 is a practical application of basic communication theory. Chapter 1 begins with an analogy between radio transmission and reception, and human communication. Then it reviews the importance of understanding the reader's point of view. Chapter 2 examines the obstacles to clear communication that are recognizable in the message to be transmitted. Chapter 3 covers structures—from the appropriateness of the presentation, through the orders of development and the outline that may be used, to the division of material into paragraphs. And Chapter 4 affords a close study of the units and combinations of language that convey the message to the reader.

Part 2 focuses on the basic principles and skills for designing and developing technical messages. Chapter 5 reviews the essentials of effective research and interpretation, concluding with a section on the possible value of imaginative as well as judicious thinking. Chapter 6 offers guidelines for handling the fundamental techniques of defining and describing the unknown and the unfamiliar. Chapter 7 covers the inclusion of helpful adjuncts to technical writing—summaries, lists, and graphic aids.

Parts 3 and 4 explain and illustrate application of the principles and skills discussed in the first two parts, and special messages prepared by technical writers. Part 3 includes the writing of letters (Chapter 8), memos and informal reports (Chapter 9), and formal reports and proposals (Chapter 10). Part 4 covers two special writing problems: articles (Chapter 11) and the job application (Chapter 12). Chapter 11 will be particularly helpful to students of technical writing who are already employed; Chapter 12, to students who are looking ahead to their first job.

Part 5, A Technical Writer's Handbook, includes sections on abbreviations, numbers, spelling, sentence unity and coherence, and diction. Also included is a list of grammatical terms to aid the student who has forgotten—or never really understood—the names that are used in discussions of grammar and functions.

Exercises for both discussion in class and writing assignments follow all chapters and the handbook. Major chapter headings are repeated in the exercises, so that teachers can readily identify exercises that are based on individual sections of the text. Most of the writing exercises are included under the heading Special Problems.

Because the basic principles of writing do not vary, this book contains ideas that were published earlier under the title *Resourceful Business Communication* (Harcourt Brace Jovanovich, 1966). However, almost all of these ideas are here presented anew and with fresh examples. I am grateful for being able to include the ideas, along with a few of the illustrations that are especially appropriate to the teaching of technical writing.

I owe much to a number of people for the completed text. I am indebted to the book's reviewers, Professor Margaret Blickle of Ohio State University and Professor Anthony Lis of the University of Oklahoma, both of whom made valuable suggestions for improving the text. I should also like to acknowledge my students at school and others for their ideas for material and for illustrations.

At The Pennsylvania State University, in addition to the encouragement of Dr. Henry W. Sams, formerly Head of the Department of English, I am grateful for many conversations with Professors Emeriti John S. Bowman, Harold F. Graves, and Edward J. Nichols; Professors Lynn Christy, A. M. I. Fiskin, Gale G. Gregory, William H. Hill, James F. Holahan, Kenneth W. Houp, S. Leonard Rubinstein, Albert Skomra, James P. Stewart, and Robert G. Weaver; and the late Professor Lyne S. S. Hoffman.

I should also like to thank my colleagues in the American Business Communication Association, including Professors E. Glenn Griffin of Purdue University and Francis W. Weeks of the University of Illinois, and my colleagues in the Society for Technical Communication, including Professors E. Rennie Charles of Ryerson Polytechnical Institute and Jay Gould of Rensselaer Polytechnic Institute.

WILLIAM A. DAMERST

CLEAR TECHNICAL REPORTS

UNDERSTANDING
CLEAR COMMUNICATION

Chapter One

THE BASIC PROCESS AND READERS' WANTS

An employer said recently that he wished he could put "MUST BE A SKILLED WRITER" in job advertisements. When he was asked why he couldn't, he said: "Can't you imagine what would happen? That line would scare applicants away!"

It is very likely that you are reading this book to further your development as a specialist. However, suppose you stopped reading at this point because you had to find a permanent job immediately. How would you feel about having to meet the requirement "MUST BE A SKILLED WRITER"? Would you be upset if the line appeared in the advertisement for a job you wanted? Would you say, perhaps, "But I never mastered the skills of writing"? Or, "I never knew writing would be that important, so I didn't work hard in English class"? Or possibly even, "I never did like English, and this requirement doesn't make me like it any better"?

Naturally, every answer suggests another to you. You probably would have many reasons for being upset—but none of them would help you get the job. Of course, you really don't have to worry. It is unlikely that an employer would require all applicants to be skilled writers.

However, employers generally do wish their workers could write better. Those who do the hiring in business, industry, and government can tell you why *they* are upset about poor writing. They readily point to losses of contracts, declines in profit, and increases in cost. They can cite specific problems in their organizations either directly or indirectly caused by poor writing.

This book begins with comments on the practical world you will work in. Before you learn anything else here, you need to know that your specialized training is only as good as the sum of its parts. It is not enough to be technically competent. In the business world, being able to do the work is merely part of the job; part consists in being able to tell others very clearly, in writing, what you have done. "Doing your work" means doing the *whole* job—not just part of it.

UNDERSTANDING THE COMMUNICATION PROCESS— THE KEY TO GOOD PRACTICAL WRITING

To write well in the practical world, you must understand *communication. Communication*, broadly defined, *is the sharing of information.*

What, exactly, do "sharing" and "information" mean? Although there are many theories of the nature of the communication process, we will use a very simple one to clarify our definition. First we will find it helpful to establish an analogy, or similarity, between radio transmission and reception of sounds, and human communication in writing.

In radio transmission, sound is encoded, that is, converted, by the transmitter at the source into clear electric signals. The signals are then sent to the destination by electromagnetic waves where they are decoded, or converted back, into sound by a receiver. When we turn on our radio, we are able to hear news broadcasts, recorded music, and other programs if the process is completed as described. Figure 1 shows this process, which may be quite familiar to many of you.

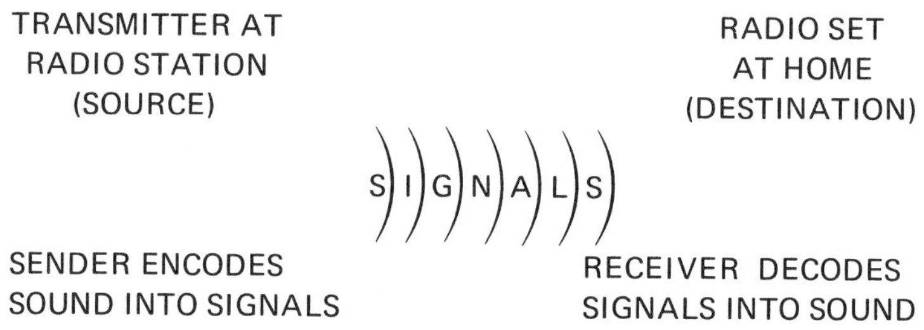

TRANSMITTER AT   RADIO SET
RADIO STATION   AT HOME
(SOURCE)   (DESTINATION)

S I G N A L S

SENDER ENCODES   RECEIVER DECODES
SOUND INTO SIGNALS   SIGNALS INTO SOUND

Figure 1

In human communication, the writer is analogous to the radio transmitter. His message, similar to sound, must be encoded into words that will be clear to his reader. The words are then sent to the reader in one of the common forms for practical writing—letter, report, or memorandum—or in another appropriate form. At destination the words are decoded by the reader into the writer's message. When the reader picks up the message, he is able to share the writer's information if the process is completed as described. Of course, the signals (words) must mean the same thing to the decoder (reader) as to the encoder (writer). Figure 2, picturing this process, will not be as familiar as the first figure, but the analogy should help.

The analogy appears to be perfect. The writer (transmitter) sends out the message (sound) in words (signals) by means of a form (waves). The reader (receiver) changes the words (signals) into the message (sound) so that he can share what is sent.

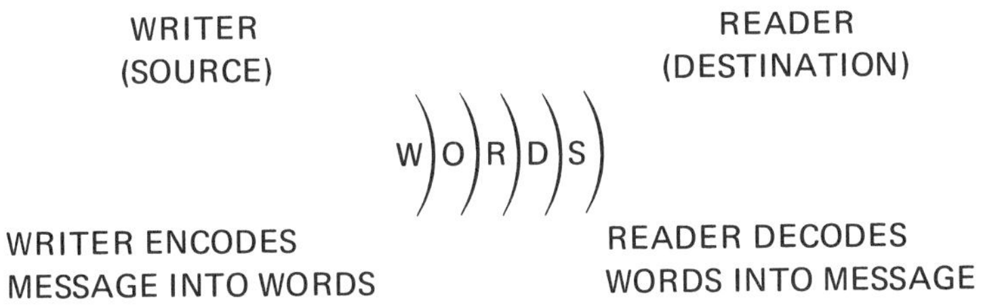

WRITER   READER
(SOURCE)   (DESTINATION)

W O R D S

WRITER ENCODES   READER DECODES
MESSAGE INTO WORDS   WORDS INTO MESSAGE

Figure 2

"OLD" TECHNICAL TERMS AND A NEW ONE— AND THEIR IMPORTANCE

Before we can go on, we must establish two other analogies: one to explain the use of "encode" and "decode"; the other to explain a related technical term, "feedback."

Encoding and Decoding

"Encode" and "decode" are helpful terms for understanding written communication because they indicate that the message, like sound, should be identical at source and destination. "Code" in encode means exactly the same as "code" in decode: what the sender "puts in" (*en*codes), the receiver "gets out" (*de*codes). (The use of a code in espionage illustrates.) Thus, no change takes place in either the sound or the message, even though neither is itself transmitted. Sound is transmitted as signals, and the writer's message is transmitted as words.

Of course, neither sound nor message is received *exactly* as it was sent. In radio, an oscillator that drifts, and amplifiers and multipliers that increase output signals too much or too little, will certainly cause distortion. And the distortion will increase when microphone and receiver also function poorly or when a disturbance such as a thunderstorm affects radio waves. Even if the transmission is electronically perfect, the fact that we at our radio hear differently than we would at the studio, indicates that the sound received is not precisely what was sent. The difference here can be compared to that between hearing a friend's voice on the telephone and hearing it in a face-to-face meeting. The two are never exactly the same.

In human communication, similar problems can arise which we will look at shortly. For now, it is enough to say that they are even more likely than problems in radio communication. All of us differ greatly in our background, knowledge, thinking, and understanding. No two people have exactly the same experience, even when they are together. Naturally, the more we differ, the more difficult it is for us to share information. Indeed, in some cases it is very possible that we will not even agree on what information—message—we are to share!

Yet, if communication is to take place, the message must go from writer to reader with as little distortion as possible. "Encode" and "decode" are terms to remember, then, if only because they serve to remind the writer of this point.

Feedback

Can there be "as little distortion as possible" when people are so different? A term made popular during the growth of automation—"feedback"—gives us hope that there can be.

In the *Funk & Wagnalls Standard College Dictionary*, feedback is defined as

the return of part of the output of a system into the input for purposes of modification and control of the output, as in electronic amplifiers, automatic machines, certain biological and psychological processes, etc.[1]

In our analogy between radio and written communications, feedback would mean that part of the signal and part of the message must return to the sender. Thus, he can learn just how well his message is being received and make adjustments where they are required. But can there be such a return? Realistically speaking, we note that in radio communication a check can be made only as far as the sound can be traced. In written communication a check can be made only before the message is sent. The sender cannot go into every home that has a radio, or into every office that is the destination for his message. Even when the sender is on the other side of a glass wall transmitting ham messages to us, or is sitting in our office when we read his message, can he be sure of our reaction?

Clearly, where human communication is involved, feedback can be only partial. One cannot check the most important element of the process—the individual who receives what is sent. Nevertheless, even partial feedback is valuable. The radio station can sample listeners, and the writer can study readers' reactions to messages like his own. Control devices can automatically correct errors in transmitting equipment. Similarly, the writer can correct his message, before sending it out, by applying what he has learned about communicating effectively.

ANTICIPATING OBSTACLES IN THE COMMUNICATION PROCESS

By encoding carefully and paying attention to feedback, the sender can avoid obstacles in both radio transmission and written communication. Naturally, our precise analogy will weaken a little when we consider obstacles. We cannot find similarities between all that can go wrong with radio broadcasting and all that

---

[1] *Funk & Wagnalls Standard College Dictionary*, New York: Harcourt Brace Jovanovich, Inc., 1963, p. 487.

can go wrong with written communication. However, we could easily find counterparts in written communication for a weakness in radio-sending equipment, a flaw in the transmission, a fluctuation in the signal, interference—such as static—with the passage of electromagnetic waves, distortion in the signal received, a dead transistor, and a listener's inability or failure to hear.

For example, the writer may not understand what his purpose is in writing a report. Perhaps he begins a section or a paragraph with one idea but suddenly and illogically shifts to another. Or, he may create static by not making important points in his sentences stand out, by leaving out one or more needed words, by using words that have more than one meaning, by using technical language that the reader does not understand, or by using so many long words together that the reader "gives up."

If we wish, therefore, we can keep the analogy in mind as we focus on obstacles in the written-communication process. To pinpoint the major obstacles, we will examine the parts of the process in four separate pairings, in the order indicated:

1. The writer and the reader (sender and receiver) in this chapter.
2. The message in the writer's mind and in the reader's mind (encoding and decoding) in Chapter 2.
3. The writer's forms for the message and the reader's acceptance of those forms (electromagnetic waves) in Chapter 3.
4. The writer's words and the reader's understanding of them (signals transmitted and signals received) in Chapter 4.

For now, we will drop the literal analogy because it has done its job. The analogy has shown us that communication is a complex matter and that the reception of the message is the most important part of the process.

OBSTACLES TO SATISFYING THE READER'S WANTS

The first thing to learn about your reader is that you don't write *at* him. You write *for* him. One authority on communication said, "Writing is for reading: writing doesn't mean anything until it is read."

Readers have definite *wants.* "Want" is a word to use and remember because it covers all three requirements of the readers you will write for in the practical world. An analysis of want will enable you to understand that in this context the word means

1. What readers lack—do not have.
2. What readers need—must have.
3. What readers desire—wish to have.

These three meanings of want represent three kinds of writer neglect of the reader: not accounting for your work, not reporting special work, and not helping colleagues to satisfy their desires.

Not Accounting for Your Work

A major responsibility you will have in your job is accounting for what you do. Your job may require that you write reports and other messages regularly—daily, weekly, monthly, or after the completion of each project. In the practical world, many relationships are carried on at a distance. If your work affects others, they will benefit by your writing. They lack what you know and will gain that knowledge only if you communicate regularly.

Of course, if you are near others, you can "talk out" an achievement, a failure, or a problem. But even then you should "put it on paper," as well. Your written message can be made a permanent record, which can then be referred to at any time.

Not Reporting Special Work

In addition to wanting reports that account for your work, the people you work with have special needs. A problem is discovered in your area of specialization, and a solution must be found. Who is better qualified than *you* to research the problem and present the solution?

Officers and others in your company continually rely on you for reports on special work. To do their jobs they need to know what your knowledge and professional analysis of particular problems can tell them. Reports, our primary concern in this book, literally mean "what are brought back"; the word *report* comes from the Latin *reportare* (re-, *back*, and portare, to *bring* or to *carry*). Knowing the original meaning of the word, you should never forget your obligation as a writer to satisfy demands when you are assigned a special project.

Not Satisfying Others' Desires

Readers' wants also include natural cravings for information. Colleagues want more than reports telling them what they should know and must know. They also want reports that satisfy their curiosities and interests.

Your final responsibility as a specialist, then, is writing reports on ideas that come to you in equipment, processes, methods, theories, etc.

Naturally, you as a writer must want to communicate. You need to share information and to satisfy others' wants, in order to be successful.

EXERCISES

INTRODUCTION

1. Suppose that you had to apply for a permanent job tomorrow. How would you feel about having to meet the requirement "MUST BE A SKILLED WRITER"?
2. What is your attitude generally toward the study and writing of English?
3. You probably have at least some knowledge of the business world—through

summer and part-time jobs and through relatives or friends who are now at work. What situations do you know about in which poor writing contributed to lost contracts, lost sales, declines in profits, increases in costs, and other problems?

4. Do you personally know anyone who is very successful in his job? Does the person whom you know write well? Is he or she an effective speaker also?

5. How much writing and speaking have you had to do in summer and part-time jobs? Were you grateful for the opportunities to write and speak in the job, or were you annoyed?

UNDERSTANDING THE COMMUNICATION PROCESS— THE KEY TO GOOD PRACTICAL WRITING

6. In this chapter, "analogy" is defined simply as "similarity." Specifically, analogy has a much more precise meaning. Look up "analogy" in your dictionary and come to class prepared to define and illustrate the precise use of the word.

7. Prepare from your own experience a list of terms that are analogous to each other. For example, there is a precise analogy between the camera and the human eye.

8. Explain briefly the analogy between radio transmission and reception of sounds, and human communication in writing. Can you show the analogy graphically?

9. Explain "encoding" and "decoding" in human communication.

10. Explain "feedback" in human communication.

ANTICIPATING OBSTACLES IN THE COMMUNICATION PROCESS

11. Tell the class about an experience in which, in writing, you created an obstacle for one or more readers. Use the terms in paragraph 2 on page 7.

12. Tell the class about an experience in which, in speaking, you created an obstacle for one or more listeners. Use the terms in paragraph 2 on page 7.

13. To what extent, if any, did you get feedback in the experience you told about for Exercise 11?

14. To what extent, if any, did you get feedback in the experience you told about for Exercise 12?

OBSTACLES TO SATISFYING THE READER'S WANTS

15. What is meant by the statement "Readers have definite *wants*"? What does "wants" in this statement mean?

16. To what extent have you satisfied a reader's wants in a particular situation? Supply details.

17. To what extent have you satisfied a listener's wants in a particular situation? Supply details.

18. Is it easier for you to satisfy a reader's wants or a listener's wants? Explain your answer.

SPECIAL PROBLEMS

19. Present a technical explanation of encoding and decoding a radio message.
20. Present a technical explanation of the feedback process in radio transmission.
21. Present your response to Exercise 19 so that any reader can understand you.
22. Prepare a brief composition (two or three paragraphs) in which you report, with details, your background for the course you are now taking. Discuss your background in your field of specialization (or in science, etc., generally) as well as your background in English.

Chapter Two

OBSTACLES TO UNDERSTANDING THE MESSAGE

In this chapter, we continue our examination of individual parts of the communication process, focusing on major obstacles to the reader's understanding of the message. These major obstacles, created by the writer while he is encoding, can be identified as: inaccuracy, unclear reason for writing, incompleteness, overwriting, lack of candor, lack of empathy, and omission of summary.

INACCURACY

Inaccuracy is a Hydra-headed monster. In Greek mythology Hydra was the many-headed serpent slain by Hercules. In the story, when Hercules cut off one head, it was replaced by two unless the wound was cauterized. The meaning here is that one error leads to another unless the writer tries to avoid all inaccuracies.

Three errors are especially "monstrous": inaccuracy in defining the subject, inaccuracy in assembling data, and inaccuracy in identifying the reader.

Inaccuracy in Defining the Subject

Inaccuracy in defining the subject means that the subject of your writing is not what is expected. For example, what your reader expects is a report on the probable outcome of tests you are making in the laboratory. You, however, send him only a description of your testing procedure. Or, in another case, your reader expects a solution to the problem of reducing lead content to produce a more nearly lead-free gasoline. Yet you report only that your department is having difficulty in completing the investigation by the deadline. You fail to note the successful results that you have achieved to date.

To share information, you must be sure that the message sent is the message expected. Looking at your work from your reader's point of view—that is, identifying his immediate wants clearly—is the only way you can be confident that your subject is "right."

Inaccuracy in Assembling Data

Being inaccurate in assembling data can mean that you have made errors during your research. Perhaps you have used the wrong test or have tested the wrong variable. Possibly the data that you plan to use are incomplete, are no longer pertinent, or are wholly inappropriate. It may be simply that you have neglected to check your figures.

In the last case, a small error can become a big one. For example, assume that you are a technician reporting the time you spent on various projects during a work week. You report ten hours' work on the gas chromatograph, but you actually spent twenty hours. Then suppose that your supervisor, estimating chromatograph time for a four-week project for a client, quotes forty hours instead of eighty in the proposal he sends. Total costs will obviously be wrong. Indeed, your laboratory may lose the contract because the client thinks the cost is too low for that part of the project to be done well.

If you are inaccurate in assembling data, you are sure to raise questions in your reader's mind about your competency. To be convinced of your competency, your reader must find your reports accurate in all respects. Therefore, you must be certain that everything "adds up right."

Inaccuracy in Identifying the Reader

If you are inaccurate in identifying your reader, the consequences may be disastrous. In today's complex organizational structures, a worker is often responsible to several superiors. Thus, when you begin your job you may find that you have to report to more than one reader. A major problem is likely to arise if you write to the wrong man.

For example, the superior to whom you decide to report may not be the one who most wants your information. Hence, he may not take any action at all. Or, if he finally realizes that your report should have gone to someone else, it may be too late to send it on. Or, he may think that he should act when he should not—and cause a problem by creating a commotion in the company over the "inefficiency" of workers. In this case you very likely would be reprimanded, and during the commotion your message might well be overlooked.

Something like this happened recently when a writer addressed a report to *all* leaders of a company. Concerned about a serious problem in his department, the worker wanted to be sure that everyone in authority knew about it. Unfortunately, his report was so written that all leaders felt they were being called upon to take action. The consequences were doubly negative: all leaders criticized the worker, and nothing was done about the problem.

Quite clearly, analyzing readers' responsibilities and wants is the way to be sure your message will reach the right person. Others can be sent copies of your message. It must be apparent, though, that the reader to whom the message is addressed is the one you are trying to reach *first.*

UNCLEAR REASON FOR WRITING

All these inaccuracies point to flaws in the writer's understanding of his responsibilities as a practical writer. His first responsibility is to share information, of course. You will remember that this obligation is to write so that, as much as possible, the message decoded is the message encoded. His second responsibility is to satisfy a particular reader's wants. You will remember that this obligation is to write so that the reader, in his turn, can do *his* job well.

A writer must do more than accurately define his subject and identify his reader. He must also indicate clearly his reason for writing—that is, his purpose. If his facts indicate that no action should be taken, he should show that the purpose of his report (memo, letter) is to present information "for information's sake." On the other hand, if his facts indicate that action should be taken, he should show what action is desired or recommended. In other words, every message should help the reader see what he should do (if anything) after receiving it. The following examples illustrate how a writer might indicate his purpose.

Stating the purpose of writing when no action is to be taken:

Here, for your information, is a complete description of the Westmoreland Company's new preheater.

Stating the purpose of writing when action is desired or recommended:

We are sending, for your approval, a list of suggestions for increasing production at the Manchester plant.

A complete example will show precisely what is required of a writer in a particular case. First we will look at a report that shares information in part only and that satisfies the reader's wants not at all:

Date: May 8, 19__

From: J. N. Stone, Chemist

To: Dr. F. G. Cramer, Director of Research

Subject: Diesel Oil 21, Sample B-109

Sample B-109 of Diesel Oil 21 has been received and tested. The sample failed to meet specifications for foaming. The addition of 0.0034% of Agent 251 makes it possible to meet these specifications.

The sample was hazy in appearance and gave off an odor of hydrogen sulfide.

From the sample a liter was taken and centrifuged until bright. The sediment accumulated, which was equivalent to 0.6 gram per liter, had an ash of 5.25%. By spectrographic analysis it was determined that the ash contained decomposed additives.

It may be that a usable product can be obtained by filtering and air blowing. The sample did not darken after being overheated.

Next we will look at a revision of this report that truly communicates:

Date: May 8, 19__

From: J. N. Stone, Chemist

To: Dr. F. G. Cramer, Director of Research

Subject: Diesel Oil 21, Sample B-109

Analysis of Sample B-109 of Diesel Oil 21, received May 1, shows that the haziness and the strong odor of hydrogen sulfide can be removed only with difficulty and at great expense. It is therefore recommended that the stored oil not be sold under the product name.

Through centrifuging, it was determined that the haziness was due to the presence of decomposed additives. Apparently water contaminated the oil, and the contamination in turn resulted in hydrolysis of Agents 726 and 583 used in the manufacture of the product. When hydrolysis occurs, Agent 726 forms a fine barium salt and Agent 583 forms hydrogen sulfide.

The sample was centrifuged extensively until bright. The sediment that accumulated, 0.6 gram per liter, had an ash of 5.25%--much too much. Overheating did not cause the separated oil to darken, but filtering to remove the sediment and air blowing to remove the odor are complex, expensive processes.

The blending of this oil with another, cruder oil will make the contamination unnoticeable, and the combined products will make the oil acceptable to all users. Thus, it is recommended that the Diesel oil in storage be mixed with No. 6 Fuel.

You will notice two accomplishments in the revision that are lacking in the original report:

1. The information in the revision tells the reader everything that he needs to know. The writer shares his knowledge as well as his experience in the laboratory.
2. The reader knows what to do next as a responsible official in the company. All his wants are satisfied.

Clearly, the writer of the revision understood exactly his responsibilities, and hence his reason for writing. The writer of the original did not.

Some of you may object to the harsh criticism of the original. Indeed, you might say, "The chemist reported what he had been trained to report—what he observed and recorded while he was carrying out the tests. Furthermore, as an objective scientist, he may feel reluctant to present a specific conclusion and recommendation. There *are* times when a definite answer cannot be given."

But note that the revision is no less factual than the original. Everything that is factual and significant is included in the revised report. The major difference is that the revised report does what its writer is obligated to do: help leaders to make decisions. It should be quite obvious to every reader that the facts presented are conclusive. Since the writer alone has the facts and is qualified to make specific recommendations because of his training, knowledge, and experience, it is his duty to communicate fully. In this case, the original writer's reason for writing should have been self-evident.

INCOMPLETENESS

When practical writing is complete, it anticipates all the questions the reader is likely to ask. You can readily appreciate the importance of completeness by comparing the original and revised reports that were just quoted. The original raises at least ten questions that demand answers:

1. Generally, what is the significance of all the facts reported?
2. Specifically, what does it mean to say, "The addition of 0.0034% of Agent 251 makes it possible to meet these specifications"?
3. How hazy is the oil?
4. Is the oil so hazy that we should not even consider trying to sell it?
5. How bad is the odor?
6. What caused the odor?
7. Is there a way to get rid of the odor?
8. What exactly is meant by "centrifuged until bright"?
9. What does the value 5.25% mean?
10. Is the writer making a valid conclusion when he writes, "It may be that a usable product can be obtained by filtering and air blowing"?

The revised report answers these questions—that is, the questions that *need* to be answered.

Getting into the habit of asking questions is the only way you can be confident that your writing will be complete. Sometimes, you can communicate with your reader before you write. Then perhaps you will learn directly what questions should be anticipated, and you can provide for precise "answers" in your report. If personal contact is impossible, you may be able to do as well by searching for questions as you review your finished report. The preceding list indicates the kinds of questions that decision makers are likely to ask.

OVERWRITING

Overwriting is wordiness in expression or overdevelopment of support. Both weaknesses are the result of a writer's fear that his work will be criticized as "incomplete." Hence, he tends to use two words in place of one and to expand his sentences into structures that are two or three times as long as they should be. Or, he includes every detail about his subject that he accumulates during his research. And in so doing, he annoys his reader, who wants to know only the essentials. In an extreme example, both weaknesses may be very much in evidence.

Wordiness alone creates a problem for the reader because he must work to understand the simple idea being expressed. The following sentence illustrates the weakness:

```
Our concern for and handling of quality control methods over
the years have been characterized by a period of extremely strict
adherence to the requirements originally set up, on the one hand,
and very indulgent observance of the requirements, for a similar
period of time, on the other.
```

The reader will be annoyed by this sentence because when he finally gets the message he realizes that all the writer had to say was

```
Our handling of quality control has been alternately very
strict and very loose.
```

In its other form, overwriting reminds the reader of a bad storyteller, who omits no detail. The amount of information included to support a general statement may be so great that the reader loses track of the writer's essential idea. This excerpt from a student report is a good example of overdevelopment (the raised numbers are footnote symbols[1]):

```
The use of DDT to treat Dutch elm disease has caused the
death of an untold number of birds. This fungus disease is
spread from elm tree to elm tree by bark beetles. In an effort
to stop the spread of the disease, infected elm trees are
sprayed with DDT in the hope that the beetles will be killed be-
fore they can transport the spores to other elms. The elm trees
are usually sprayed with 2 to 5 pounds per 50-foot tree, and if
elms are plentiful in the area this amount of DDT results in up
to 23 pounds per acre.[1] Reports appearing in Audubon Magazine
state that five pounds per acre kills all birds living there.[2]
A case of this occurred on a midwestern state university campus
in 1954.[3] A graduate student happened to be studying robin popu-
lations on the campus the year after a program had been started
to spray elm trees for disease. When the robins returned the
next spring, many suddenly started to die, showing the character-
istic signs of DDT poisoning: shaking, loss of balance, and con-
```

---

[1]Footnoting is covered in Chapter 5, pages 79–81.

vulsions. The robins had eaten earthworms containing a high concentration of DDT. Earthworms themselves are not affected by DDT,
but they store it in their tissues. "As few as 11 large earthworms can transfer a lethal dose of DDT to a robin. And 11 worms
form a small part of a day's rations to a bird that eats 10 to
12 earthworms in as many minutes."[4]

This paragraph is an example not only of overwriting but also of poor paragraphing.[2] We can almost hear the original reader saying, "Get on with it"—
especially after reading about the beetles in sentences 2, 3, and 4.

Perhaps it would be oversimplifying to reduce the preceding paragraph to
these four sentences:

The use of DDT to treat Dutch elm disease has caused the
death of an untold number of birds. Of necessity, elms are
sprayed with as much as 5 pounds per 50-foot tree[1]—an amount
that, diffused over an acre, would kill all birds living there.[2]
As discovered at a midwestern state university, drippings of the
spray saturate the soil and are assimilated by earthworms feed-
ing underground beneath the trees. Although the worms are un-
affected, birds that eat the worms eventually die, showing first
all the signs of DDT poisoning: shaking, loss of balance, and
convulsions.[3]

However, these four sentences report all that *needs* to be said.

What we have been talking about here is the need for conciseness in technical writing. The careful selection and organization of details into a unified,
coherent, and emphatic presentation enable one to achieve conciseness. Thus
"defined," this quality is always desirable because it enables the writer to present the reader with only the essential ideas.

To develop and master the quality of conciseness in your own writing, practice the writing of summaries. Summary writing is described briefly at the end of
this chapter and more fully in Chapter 7.

LACK OF CANDOR

Lack of candor is, or may seem to be, a major obstacle because the person who
thinks you are less than honest will undoubtedly put little faith in what you
write. Of course, the reader will realize that inaccuracies may appear in your
writing. Indeed, he may overlook them if they are not vital and if you do not
make the same kind of mistake continually. He will do the same about your failures to make clear your purpose and to make your writing complete and concise,
if he can see that they are "honest" failures.

What the reader will not tolerate, however, is bias—prejudice—in your writ-

---

[2]See Chapter 3, pages 34–38, for a detailed analysis of paragraphing.

ing. No decision maker can take an action with confidence if he recognizes in your writing a desire to satisfy a personal interest. Good report writing is objective; it is impartial. A report cannot be considered objective if it does not take into consideration all facts and all conclusions, that are relevant. For example, such a report might praise the virtues of one generator without evaluating other generators that perform as well.

This discussion of candor is not intended to indicate that conclusions and strong recommendations should not be made if all the facts point to them. It *is intended* to indicate that anything less than a thorough, objective report will almost certainly lack the candor the reader needs to find if he is to make a decision with confidence.

LACK OF EMPATHY

"Lack of empathy" may be less familiar than the other obstacles we have discussed. Empathy means appreciative understanding of the reader's point of view. Literally, it means "a feeling in," trying to think and feel as the reader thinks and feels. Being empathic with your reader presents you with no small challenge. Indeed, you know from experience that no two people think and feel the same about a subject. The most you can reasonably hope for is to have some success in writing emphatically.

Lack of empathy in report writing is apparent when one or more of the obstacles just discussed are obvious. It is difficult to achieve empathy if your report contains inaccuracies, seems purposeless, is incomplete, is overwritten, or reveals your bias. These weaknesses clearly indicate that you really are not looking at your subject with the reader's point of view in mind.

An equally apparent lack of empathy is obvious when you are indifferent—perhaps even discourteous—in your writing. Treating your subject and material with a casualness and a lack of concern for satisfying wants will impress no one. Your reader will quickly suspect that you do not take your job, or at least this part of your job, seriously. Indeed, your apathy may lead him to believe that you are neglecting your responsibility.

If you are discourteous as well as indifferent, your reader will be doubly distressed. Discourtesy may seem impossible in report writing, because you have been trained to be objective. Consider this all-too-common situation, however. Your reader has an idea—let's say it is a solution to a problem—which he shares when he asks you to investigate the problem. In your study, though, you find a much better solution. Then, in your report you ignore—or virtually ignore—his solution and focus on the qualities of your own. Surely he will wonder why you have passed over his idea, and almost as surely he will be offended.

In a case like this, by writing a complete report—evaluating all practical, possible solutions, based on objective criteria, of course—you demonstrate empathy.

No realistic person will insist on using his solution if the facts indicate that it is much less effective than another.

Sometimes you can demonstrate empathy by including concession and conciliation in your report. The first two sentences in the following excerpt illustrate concession.

    The Blaisdell Method has a number of fine features. It would
    offer us . . . [list of features shown by factual support to be
    desirable]. However, it would have the following limitations if
    it were adopted here: [list of limitations accompanied by strong
    factual support].

Conciliation in report writing might appear in this way:

    The Blaisdell Method would work out well in our plant if our
    operation were less complex. However, . . . [list of reasons,
    based on facts, for not adopting the method].

These techniques help you to be courteous in reports because they have a positive quality. They make the reader feel that his idea made a contribution to the study, if only to ensure more complete coverage of the subject. Thus he can make his decision with greater confidence.

Lack of empathy in letter writing is even more noticeable. A letter is usually a more personal medium of communication than a report. It is characteristically shorter, more generalized, and less detailed. Often its function is to fill in the gaps during an investigation to be covered by a long report. As a "catch-all" kind of writing, therefore, a letter invites a close, friendly, and not always solely objective relationship between writer and reader.

For these reasons, letter writing may be a real challenge for someone schooled in report writing. Because letters are short, the words in them frequently carry a heavy burden and one misunderstood word can mean the end of a pleasant relationship. Take the word "want," for example, which we have shown to carry a number of possible meanings. If you use it in a letter to indicate "desire" ("We want your report to be as thorough as possible"), the reader may think you mean "demand." Your use is casual, but the reader may see it as a threat. If he does, several more letters and even a telephone call may be needed to understand (and accept) your use.

As a personal medium of communication, a letter requires the use of civilities (courtesy words and phrases), which are seldom used in objective reports. Some examples are the underlined phrases in the following excerpts.

    Thank you for your letter of June 19 asking us for more de-
    tails about the scrubbing process used in our plant.

    In the past few months we have heard a great deal about the
    fine quality of your Series Y gas manometer.

    We will appreciate your sending us a copy of your specifica-
    tions by August 18, so that we can work out details of the con-
    tract to your complete satisfaction.

Also, because letters are often supplemented by meetings, it is common for first names to be used in correspondence:

Dear Harry:

We sincerely welcome your looking over the details of the installation, Bob

As indicated by these illustrations, a different tone is often found in letters than in reports. Empathy is required in both kinds of writing, of course, but you must worry about it much more in letters. You cannot seem to be grudging, negative, or hostile—tones that are obvious (and undesirable) in these sentences:

After reviewing your letter of June 19, we have decided to send you more details of the scrubbing process used in our plant.

Although at present we have no intention of replacing the gas manometers in our laboratory, we understand that your Series Y manometer is a new model on the market.

We must have a copy of your specifications by August 18 or we will be unable to work out details of the contract by September 1, the date your office and ours firmly agreed on last month.

The excerpts just quoted not only will irritate readers but may even cause a loss of business and an end to relationships!

Always remember that in both reports and letters, you are writing to other human beings. Remember, too, that empathy in your writing will help you get your message across.

OMISSION OF SUMMARY

The absence of a summary is covered last in this chapter because the summary of a report is not really part of the report. A true summary is the report in brief —a capsule statement of the whole. Its importance in today's technical reports is illustrated by the often-heard plea of the executive to "put it on a page." What he means is that as far as he as a decision maker is concerned, a page of writing should be enough for him to take the action recommended or indicated by the whole report.

Usually, details presented in reports are so numerous that the length of a message requires several pages. Executives and other leaders, however, rarely have time to read reports from beginning to end. Hence they ask writers to add to their reports a complete summary of the entire message, with emphasis on the conclusions drawn and the recommendations made.

No leader welcomes obstacles to understanding the message (inaccuracy especially, of course). But if the obstacle is not serious, a superior probably will be satisfied as long as the report includes a good summary. By mastering the technique of summarizing, therefore, you can make amends at least partially if you are having difficulty in eliminating the other obstacles.

The technique of good summary writing is covered in detail in Chapter 7. However, an example of an effective summary here will help you to anticipate the extended coverage. Compare the following with the revised report on page 14:

```
    Because it is badly contaminated, the Diesel Oil 21 in stor-
age should be mixed with, say, No. 6 Fuel, where the contamina-
tion will not be noticeable.
```

This summary, less than 15 percent of the length of the revised report, would tell the reader everything he needed to know. And it would take him only a few seconds to read it.

EXERCISES

INACCURACY

1. What inaccuracies, if any, can you find in the following:
   A writer's title for his report:
   "Methods Proposed for Treating Radioactive Waste from Research Reactors"
   The same author's writing of the title for his abstract (summary) of the report, the abstract having been written for separate distribution to other interested employees of the company:
   "Proposed Plant Design for the Treatment of Radioactive Waste Effluent"
2. What inaccuracies, if any, do you find in the report that is set off below this paragraph? The report writer is responding to the following note from L. R. Jones, his department manager: "Your supervisor, T. D. Crane, informed me today that you have not yet reported the results of your efforts to reformulate Compound J-2. As you know, the government's approval of Compound J-2—as that product is now formulated—expires in three weeks, on June 30. Therefore, we must know at once whether or not to withdraw this compound from the government's Approved Products List." The report writer's answer, submitted "at once," is as follows:

```
Dear Mr. Crane:

    To date we have formulated Compound J-2 twice, with varying
results. We expect to complete another reformulation by June 26
and will have our final report in your hands by the June 30
deadline.
```

3. What inaccuracies, if any, can you find in the following table, which is taken from a report entitled "Evaluation of Three Processes for Removing Sulfur Dioxide in the Ayer Plant's Combustion Gases"?

TABLE 5

ESTIMATED DIRECT ANNUAL OPERATING COST OF THE
ALKALIZED ALUMINA PROCESS

Direct Costs:			
Raw material			
Absorbent make-up		$ 910,800	
Coal		479,900	
Power		99,200	
Heat (Includes credit of $ 314,200)		125,600	
Water		28,700	
			$ 1,958,400
Direct labor			
72-man-hr/day		79,500	
Supervision		11,900	
			91,400
Plant maintenance			
23 men @ $ 6,600		151,800	
Supervision		30,400	
Material		75,900	
			268,100
Payroll overhead			50,600
Operating supplies			51,600
	Total Direct Cost		2,420,100
Indirect Costs:			200,500
Total Capital Charges			1,210,600
	Gross Operating Cost		3,841,200
Credit (Recovery of sulfur)			-1,315,800
	Net Operating Cost		$ 2,425,400

UNCLEAR REASON FOR WRITING

4. Do the two titles in Exercise 2 indicate that the report writer had a clear reason for writing? Discuss your answer.
5. Does the writer of the report quoted in Exercise 3 indicate that he had a clear reason for writing? Discuss your answer.
6. Do you think that the writer of the following report had a clear reason for writing? Discuss your answer.

We have completed our inspection of current inventories of products that we have been using for packaging our finished goods. You will note that all the figures have been adjusted to the close of business yesterday.

INCOMPLETENESS

7. In your opinion, is the report quoted in Exercise 3 complete? Discuss your answer.

8. In your opinion, is the report quoted in Exercise 6 complete? Discuss your answer.

OVERWRITING

9. To what extent is overwriting apparent in the following excerpt from a report?

> On March 9, in response to our request of March 1 sent to Allied Chemical Company, New York, we received at the laboratory two eight-ounce bottles, one containing a sample of Compound 17-25 and the other a sample of Stabilizer 11. Both of these products are antioxidants made by Allied. We opened the samples at once for the specific purpose of testing and evaluating them for use as possible turbine-oil antioxidants in our own finished products.

LACK OF CANDOR

10. To what extent is lack of candor apparent in the following report? The report was written by a new supervisor, on the job for six weeks, at a manufacturing plant.

> I recommend that the heater in No. 4 Boilerhouse be replaced at once. During the quarterly inspection yesterday, we found that the hood was thin in several places and that the end plates supporting the hood were badly corroded.
>
> I understand that the practice here has been to patch hoods and replace end plates during shutdown of the boilerhouses for heater inspections. The men tell me that these repairs are neither expensive nor time consuming. However, the men also tell me that these repairs are only temporary, lasting six months or so. The only solution that we can depend on for the next year, it seems to me, is to replace the entire heater now.

LACK OF EMPATHY

11. The following paragraph is the beginning of a report written by a field agent for his employer, the owner of a large dairy farm. To what extent do you find that the paragraph lacks empathy?

> Presently, the only measure being taken to prevent pasture bloat in the herds at Green Valley Farms is typical of an antiquated form of management. Not only is this management resulting in the deaths of cows, but also it is limiting the full use of the alfalfa pastures and therefore depriving the herds of a high-quality forage.

12. To what extent do the following sentences, taken from letters, show a lack of empathy for the reader?

a. We are unable to send you our bid because you failed to include precise specifications in your letter.
b. You must sign the contract on the line indicated and return one copy to us within seven days.
c. Our company cannot fill your order for valves because you neglected to sign it.

SPECIAL PROBLEMS

13. Revise the writing in Exercises 1, 2, 3, 6, 9, 10, 11, and 12 to eliminate the obstacles they now present to communication. For each exercise assigned, write out any assumptions you must make and include these with your revision.
14. Practice summary writing by reducing the excerpt quoted in Exercise 9 to a single sentence.
15. Revise a writing assignment that has been graded and returned to you, to eliminate any obstacles covered in this chapter. Also prepare a brief report explaining the changes you have made and stating why you have made them.
16. Write a report of at least 150 words, identifying the obstacles to communication (covered in this chapter) that you find in the following student report. The report was written for a reader who had little or no knowledge of the textile industry.

```
        DEVELOPMENTS IN THE TEXTILE INDUSTRY--IMPLICATIONS
                        FOR THE CONSUMER

     Textile products are consumed throughout the world. Per cap-
ita fiber consumption is rated second to food as an indication
of the standard of living in developed and developing areas of
the world. According to the United Nations, annual per capita
fiber consumption ranges from 2.5 kilograms in developing
countries to over 12 kilograms in the developed countries. The
man-made fibers account for approximately 50% of the total
weight consumed, and over 50% of the dollar value. The consump-
tion of the natural fibers on the basis of weight, remains al-
most constant from year to year, whereas the consumption of the
man-made fibers steadily increases. Whether the fibers are natu-
ral or man-made, an understanding of no-iron finishes, laminated
fabrics, and knits is important to the consumer. The three fabric
treatments are indicative of present consumer acceptance. They
may be classified as a time-saving, easy-care group.
     The no-iron finishes are achieved by a variety of treatments.
It is predicted that in the future, 100% of many fabrics will not
require ironing. The tumble dryer does the ironing.
     A laminated fabric is composed of two or more layers of
cloth joined with glue, resin, or other adhesive by heating. In
this way, "linings" are applied to a fabric during manufacture.
Though there are flaws in this process, it can be used success-
fully to save time and labor.
```

A knit is a textile fabric formed by a yarn connected in loops. While knits are not new, it has only been in recent years that mechanical knitters have produced large quantities of very satisfactory knits. These comfortable, easy-care knit fabrics are expected to take larger shares of the market.

No-iron finishes, laminated fabrics, and knits are expected to make up an increasingly large percentage of the textile products on the market.

Chapter Three

OBSTACLES TO ACCEPTING THE FORMS

As we continue our analysis of written communication, the temptation is great to return for a moment to the extended analogy in Chapter 1. In radio transmission, electromagnetic waves in the air are the means by which sounds are transported to the receiver. In human communication, many carriers are the means by which messages are transported to the reader. Like electromagnetic waves, we should note in passing, these carriers—or forms—differ.

The most common carriers are letters, reports, and memorandums. In this chapter we first examine practical writing as the message is conveyed by these carriers. We then consider certain techniques used in preparing these carriers effectively. You will notice that extensive explanation is included in each section of the chapter. Knowing what the effective handling of a form requires will help you to better understand the obstacles—and thus to avoid them when you write.

OVERALL FORMS

Although variations may be appropriate, definite time-honored structures, or forms, exist for reports, letters, memorandums, and other carriers of practical messages. You need to learn the stereotypes of the overall forms because they have been used successfully for a long time. Since the use of forms is conventional, we expect a certain one every time we receive a message.

Letters, Reports, and Memorandums

For example, think about the form of the business letter you received yesterday, the day before, or last week. From previous experience, when you opened the envelope you expected to find something like the following:

```
                    L E T T E R H E A D


                    Month Day, Year


Mr. Your Name
Your Street Address
Your City, State Zip Code

Dear Mr. Your Name:

     Paragraph 1 begins here......................
.......................ends here.

     Paragraph 2 begins here......................
.................................................
......ends here.

     Paragraph 3 begins here......................
.................................ends here.

                    Sincerely yours,


                    Writer's Name
WN:sn
```

And very likely that letter's form was close to what you had expected. (This form—called semiblock because paragraphs are indented—is shown here as prescribed by most authorities, and with recommended margins and spacing between elements. It is the form used for business letters by companies, institutions, and agencies.)

Or, consider a typical company's formal-report form. Perhaps the report is presented within a binder, having a title page, table of contents, list of illustrations, text with headings, and appendix. (See the example on pages 221-35 in Chapter 10.)

Or, perhaps you have seen a company's printed form for informal reports and memorandums (intracompany letters, frequently called "memos"). Conventionally these look like the following illustration, whose bottom is cut off because the printing appears only at the top:

```
                    C O M P A N Y   N A M E

     DATE     Month Day, Year

     FROM     Writer's Name, Title

     TO       Reader's Name, Title

     SUBJECT  Title for the Message
```

This form for interoffice correspondence is usually filled in without the salutation (Dear Mr. _____:) and the complimentary close (Sincerely yours,) that are standard in most letters.

Uses of Forms

Most of the time the form used is appropriate to the situation. Companies use letter form for messages to other companies and to individuals in the public at large. Letters are written to inform (notify of a change in specifications, describe a new product, etc.), to request (ask for information, extend an invitation to bid, etc.), to solicit (try to persuade a customer to buy instruments, chemicals, etc.), and to reply (send information requested, refuse a request to adapt a patented process, etc.).

Companies use memo (interoffice correspondence) form for messages to workers within their organization. Memos are written for such purposes as informing workers of a meeting, requesting a laboratory test, soliciting employees to conserve in using supplies, and replying to a worker's request to change his vacation period.

Companies use the same form for informal reports to workers and leaders. Informal reports are written to account for regular work done, to share the results of a special study, and to supply information for information's sake alone. Formal reports serve the same purposes; however, they are carefully prepared so that company officers, clients, investigating agencies, and others will have messages in a form they can keep as permanent records.

Obstacles to the Reader

Letters, memos, and reports are so common that the misuse of one might distract the reader or even affect his decoding of the message. For example, the presentation of the message might be conventional but the layout (no margins, poor centering on the page) distracting. Or, the typing might be poor (strikeovers, messy erasures) or startling (too many or not enough initial-capitals). Some company leaders insist on neat, clean presentations by their writers and typists and are disturbed by poor presentations they receive from other organizations and individuals.

The most common obstacle to accepting particular messages is the use of unexpected forms. For example a client would be intrigued if he received a business letter in memo form. Or, an executive would be distressed if he were sent a formal report in letter form.

The misuse of forms can adversely affect the message in an even more significant way. For example, consider what happened when the main office of a multiplant company learned one day that a necessary raw material was no longer available from the supplier. A report was therefore written and mailed to all company plants. However, the officer who wrote the report failed to check on the amount of raw material in stock at each plant. All but one had enough to continue in operation until another supplier could be located and a new supply shipped to each plant. However, the one that ran out of stock had to stop production for two days, and as a result thousands of dollars were lost.

The two days were the time it took for the report to reach that plant! Obviously, a telephone call to each plant would have been the best form in this case. The office would have known about the one plant's problem at once and could have made arrangements the same day for a shipment from another supplier to arrive in time.

Clearly, when you think about using a conventional form for a message, you must be sure that the form is "right" for the situation.

ORDERS OF DEVELOPMENT

Commonly Used Orders

Next we look within the form of a message at the orders of development used by the writer. The most commonly used orders of development for a message are time, space, familiar to unfamiliar, comparison, simple to complex, contrast, cause to effect, effect to cause, generalization to details, and details to generalization.

Time order is first-to-last order, or the order of sequence. The message is presented in chronological order. Time order is used for describing a process, narrating a meeting's progress from beginning to end, reporting laboratory experimentation, etc.

Space order distinguishes physical locations and relationships. It is used to report the layout of buildings and facilities at a plant site, to describe the positions and relationships of the parts of a machine, etc.

Familiar-to-unfamiliar order takes the reader from what he knows and understands to what is similar but not known and not understood. It is used to explain, for example, the working of a complex machine by explanation of a machine whose similar principle of operation the reader already knows. You will recognize the usefulness of analogy in developments by this order.

Comparison order is similar to familiar-to-unfamiliar order except that all processes, instruments, etc., being compared are well known and understood. Two fly wheels, built by different manufacturers but for the same use, are typical subjects for comparison.

Simple-to-complex order resembles familiar-to-unfamiliar order but differs in that comparison is not used. The writer proceeds from elementary to more difficult material. For example, to explain a new theory an engineer begins with an explanation of principles of physics that the reader needs to understand the theory.

Contrast order is the opposite of comparison order. Instead of showing likenesses, the writer shows differences. This order is common in writing undertaken to distinguish theories, methods, apparatus, etc., that serve the same purpose but have differences. It is used extensively in evaluations made to select one process, machine, etc., over one or more other processes, etc.

Cause-to-effect order begins with causes and ends with results. This order is needed in the analysis of how a problem came into being.

Effect-to-cause order begins with a statement and review of results (for example, the writer presents a problem and its effects) and ends with an analysis of the factors that brought about the effects.

Generalization-to-details order is the order of conclusion, summary, or explanation first and specifics last. What follows the generalization supports, adds up to, or substantiates the general statement. Reports beginning with conclusions, a statement of action taken, etc., are developed largely in this order.

Details-to-generalization order is the opposite: specifics are presented first, and the conclusion or summarizing or explanatory statement appears at the end.

Uses of Orders of Development

Rarely is a single order of development used exclusively. For example, in most complete studies, a good laboratory report is developed by a number of orders. After an introduction clearly indicating the purpose and scope (that is, the extent of coverage), the writer presents something like the following. (The additions in parentheses name the types of orders used successively.)

- General conclusion(s) and recommendation(s), if expected (generalization(s))
- Review of previous work, or history, if appropriate (time)
- Lists of apparatus and materials, if needed (generalization to details and perhaps contrast)
- List of procedures followed, if needed (time, and perhaps space and contrast or comparison)
- Results of work covered by the report (details, perhaps with comparison or contrast)
- Discussion of results (generalization and probably some details)
- Conclusions (generalizations)
- Explanations (cause to effect) (may be included under "Discussion")
- Recommendations, if appropriate, perhaps including specific suggestions for continuing the work (generalization-to-details)

The specimen on pages 197-204 includes most of these orders. You will notice there that comparison is used throughout, along with contrast where appropriate, to reinforce the overall conclusion and recommendation. In that case, there is precedence for the actions to be taken.

Obstacles to the Reader

Obstacles to the reader's acceptance of the orders of development are present when orders are expected but not used. For example, the writer of the report just cited was apparently trained to present all relevant background information. In this case, his focus is on history, and hence he used time order. His reader would have been irritated if the history (and the comparison in the sections following) had been left out.

In another reporting situation, an unskilled writer presenting a theory might confusingly mix one order with another. If the theory was difficult to understand, he might try to use analogy (familiar-to-unfamiliar order). But because he was unskilled he would probably use generalization-to-details at the same time, because that is the characteristic order for the explanation of a theory. His reader would almost surely be bewildered.

The laboratory report on pages 197-204 illustrates another point about orders of development. Note that the report begins with "Conclusion and Recommendation." Obviously, his reader wants first the information he needs in order to make a decision. If the information were not where it is—if he had to look through the report for it—he probably would be irritated.

We have noted that a writer should consider using an unconventional form whenever he can justify it. The same can be said about using an unexpected order of development in a special situation. For example, in a research and development company several very serious accidents occurred in a laboratory within the span of a week. In each case, carelessness was the cause. The project leader in charge of the laboratory thought about writing and posting a warning against carelessness. He then remembered that he had done so before and that the memo had done little good. Instead, he tried another form—talking to the workers—but that seemed to help little: another serious accident occurred almost immediately afterward.

Then the project leader had a brainstorm. He wrote a memo about the accident that had just happened. But this time he used time and space order to cover the passing of time and the movements of the worker before and through the tragedy. Laboratory employees read the memo—and more than once. The accidents stopped.

OUTLINES AS HEADINGS

Essentially, what we are pointing to in discussing orders of development is the structure, or outline, of a message. In a finished report or memo (sometimes in a letter), the outline appears in the headings used to introduce each section. Where they are appropriate, headings can be very helpful to both writer and reader. To the writer, a heading indicates that what it says—nothing more and nothing less—is included in the section following. To the reader the message becomes more acceptable if there are headings, because he can read it easily. Indeed, if he is interrupted he can effortlessly resume reading by finding the section he was reading. Headings are like maps because they show you where you are "going."

For an illustration, look again at the laboratory report on pages 197-204. In sequence, the headings in that report are

Conclusion and Recommendation
Introduction
Comparison of Laboratory Results
Causes of Oil Deterioration
Steps Taken to Save Oil
Specific Recommendations for Changing the Oil

We find that the headings in that report are logical. Thus we can assume that the writer used a logical outline to plan the report.

Logical Outlines

The headings just quoted were in fact the heads used by the writer in making his outline. These are called *topic* heads. If they were elaborated into complete sentences, they would be called *sentence* heads. Which of the two you use in planning for writing matters little, for both really say the same thing. "Comparison of Laboratory Results," for example, becomes, say, "Following is a comparison of laboratory results" when it is expanded into a sentence.[1]

The headings, or heads in outlining, of the laboratory report cited are first-degree heads. First-degree heads indicate that the ideas they contain are of primary importance. Second-degree heads are next in importance; third-degree, next, and so on. The writer of the laboratory report apparently had no use for second- and third-degree heads. Usually, however, a writer includes them to ensure complete coverage of his material. Characteristically, an outline with several degrees of heads looks like the following specimen:

I. First-degree head (no other head is more important)
  A. Second-degree head (less important)
  B. Ditto
    1. Third-degree head (less important than B)
    2. Ditto
II. First-degree head, etc.

This structure, which is called a *number-letter* outline, probably is familiar to you. It is widely taught in theme-writing courses at all school levels. In technical writing, however, many writers use a slightly different structure, although the logic behind it is the same. This structure, called a *number-decimal* outline, looks like this:

1.0 First-degree head
  1.1 Second-degree head
  1.2 Ditto

---

[1] Of course, in the finished report the headings would be topic, not sentence, heads.

      1.3.1 Third-degree head

      1.3.2 Ditto

  2.0 First-degree head, etc.

In many technical reports, the numbers and the decimals are included in the final typed draft. Without change, the outline head becomes the report heading.

Illogical Outlines

The advantage of the number-decimal outline is that its user is continually aware that his planning for writing must be logical and complete. Since the numbers always begin the same, he can check, as he plans, to see if what he has for 1.3.1, say, really belongs under 1.0. If it does not he will realize his error and move the head. When the outline is complete, he can also see if *all* his heads, taken together, make up the total coverage indicated by the title of the report. one might almost literally say that he can "add up" the numbers to get his total. If his entries do not "add up," the outline is illogical.

Other kinds of illogic in outlining are making a second-degree head a single entry and cross-classifying degrees of heads. A single division under a head, for example,

1.0
<u>1.1</u>
2.0, etc.

is illogical because a division always creates at least *two* parts. Cross-classifying degrees of heads, for example,

1.0
1.1
2.1.1

is illogical because the entries obviously are *not equal* in importance. In figurative language, we can restate these points as follows: If you cut an apple in half, you must have two parts; and if you have apples, tangerines, and cherries together, you have more than apples.

Although you cannot change the logical structure of an outline, you can vary the arrangement of heads whenever the stereotype is not appropriate. For example, rather than follow the conventional order of "Introduction, Apparatus and Materials, Procedure, Results, etc.," for a laboratory report, you may find in a particular case that your report will be more meaningful to the reader if the order is "Introduction, Conclusions." In other words, your reader may not care to know what you used, what you did, and what you found out in detail, except in an appendix. All that he may want in the basic report is a reminder of the assignment he made, and your conclusions.

Again for illustration we can turn to the laboratory report on pages 197-204.

Suppose the reader of the report had asked only for answers to two questions: "Can the oil's life be extended if we put in an additive, or is it so far gone that it *must* be changed?" Conceivably, the basic report could consist of a short introduction (to remind the reader of his questions) and then just four words: "No and yes, respectively." (An appendix could be used for everything else in the original report, for tax data and other purposes. Presumably, the decision maker —knowing of your competency—would not read the history, results, etc., himself.)

PARAGRAPHS

A paragraph is a whole composition in miniature. That is, like a composition a paragraph is complete in itself. Paragraphs are used to show how the writer divides his ideas: he uses a new paragraph for each idea that needs to be developed fully.

Qualities of Good Paragraphs

As writing complete in itself, a paragraph should have all the qualities of a composition. Only one major idea should be covered; each sentence should logically follow the preceding; and important points should be shown to be significant. We call these qualities *unity, coherence,* and *emphasis.*

To ensure unity, the writer should have a *topic sentence*—one that identifies the subject and anticipates what follows, or one that concludes, summarizes, or explains what has preceded. Regardless of what other order or orders of development may also be used (contrast and comparison, for example), a good paragraph with a topic sentence is developed either by generalization to details or by details to generalization.

To ensure coherence, the writer should develop the paragraph logically—that is, in an arrangement that the reader can follow easily and with understanding. Naturally, order of development plays a role, as does transition (words and phrases that link key ideas). To illustrate, let's look at the first three sentences of a paragraph developed by generalization to details and by contrast:

```
    Although more expensive than the natural draft tower, the
mechanical draft tower has several advantages. It is more com-
pact and therefore requires less ground area. It is also movable,
whereas the natural draft tower, because of its heavy concrete
shell, is stationary.  . . .
```

Note how transition "works" together with the order of development by contrast, to give this paragraph coherence. The pronoun "it" obviously refers to the mechanical draft tower throughout. The details to support "several advantages" are clear-cut: "more compact" and "also movable." And finally, the connectives "Although," "and therefore," and "whereas" provide important links. "Although" must look back to the preceding paragraph (on costs), and "and therefore" and "whereas" clearly link the last half of their sentences with the first half to build the development by contrast.

Finally, to ensure emphasis, the writer should structure sentences so that the reader's eye will be drawn to the key ideas. What the writer of the preceding paragraph unmistakably wishes to emphasize is starkly apparent in the phrases "has several advantages," "is more compact," and "is also movable."

Weaknesses of Poor Paragraphing

All good writing can easily be analyzed as illustrated above. At times, you will come upon paragraphs that seem to defy analysis. If you examine them for unity, coherence, and emphasis, however, you should be able to see what went wrong in the writer's development. The following paragraph, under the heading "History" in a report entitled *The Problem of Burning Coal-Refuse Banks*, enables us to illustrate:

> With the "room and pillar" system of coal mining, undersized coal and roof rock were thrown back into the old workings; no refuse piles outside the mine were formed. But the onset of machine mining brought ungraded coal to the surface, and eight percent of the tonnage produced was then discarded on piles. Present methods of the mechanical loader bring up to thirty-three percent of the tonnage as refuse.[3] During the no-regulations period mining companies disposed of their wastes in the most economical way. Often the refuse was actually dumped in conical piles to encourage burning for shrinkage of the area needed. At those times burning refuse piles were considered a necessary, ugly, fuming evil whose extinguishment was almost impossible or too expensive. The Allegheny Court of Common Pleas in 1935 stated "that there is no feasible method of operating a coal mine without a gob pile on the surface, as no use has ever been found for this troublesome by-product of mining; that every large coal mine has a large gob pile close to its tipple; that sooner or later these piles all ignite through spontaneous combustion; that practically every large mine in Western Pennsylvania has a burning gob pile, and that there is no known means of averting such a fire."[4] People resigned themselves to the offensive sight and smell of the gob piles.

This paragraph seems to have several virtues, but when we scrutinize we find that they are only near-virtues. First, it is apparent that the ideas are all related: the subject of the paragraph is coal-refuse piles. But then we see that there is a turn in thought at the end of sentence 3; the writer stops talking about the contrast between "no refuse piles" long ago and "thirty-three percent ... refuse." Suddenly his subject is disposing of wastes "in the most economical way." And in the next sentence, which straddles "disposed of their wastes" and "dumped ... to encourage burning," he begins a third subject, the "necessary, ugly, fuming evil." Obviously, the paragraph violates unity.

Of course, we should have seen a clue in the lack of unity in just looking at the paragraph. It is a formidable one; indeed, it is about 235 words long! Writing long paragraphs that are unified is difficult. (The diversity of details we must include is itself a deterrent.) Thus, shorter paragraphs are preferred by editors and executives—the people who will "count" in your writing career. A range of 100 to 125 words is considered "maximum" (about the length of each of the last two paragraphs of your book).

Second, there appears to be a fair amount of coherence. We note the "But" at the beginning of sentence 2. We also note the contrasts—between "no refuse piles were formed" in sentence 1 and "eight percent . . . was then discarded" in sentence 2, and between that idea and "thirty-three percent" in sentence 3. Then we note the close tie-in of sentences 4 and 5: "disposed . . . in the most economical way" is followed by "burning for shrinkage of the area." Finally, we observe that sentence 6 ("At those times . . . ."), along with the long quotation, anticipates effectively the resignation of the people affected, the idea of the last sentence. But if there is no unity, there can be no overall coherence.

Third, we can see some emphasis in the paragraph. The comparative size of refuse piles (from no piles at all to fairly large ones) is stressed at the beginning. And the idea in sentence 6 emphatically sets the tone of despair that marks the last "half." (To be sure, the long quotation is an intrusion from "outside"; the writer actually only contributes two sentences to develop an idea that takes up more than half the paragraph! In this part, emphasis is not misplaced, for the whole paragraph has—however ineffectively—been building up to the point that people despair. But since there is neither unity nor coherence, there can be no overall emphasis.

The paragraph, therefore, is a poor one. It *could* be made a good one easily if (1) the first three sentences were restructured to make the contrast stronger and if a topic sentence were placed at the very beginning; (2) the idea of the fourth sentence were made subordinate to that of the fifth, which begins to introduce the last half of the original paragraph; and the long quotation were reduced in length—to no more than the last twenty-five words—or, better, summarized and integrated more effectively with the sentence preceding and the sentence following. Then there would be *two* paragraphs, of course, the first ending after "thirty-three percent of the tonnage as refuse."

Equally as questionable as the writing of the extremely long paragraph is the writing of the extremely short one. Suppose the author of the long paragraph quoted had divided his original writing logically. He would have one paragraph for the first three sentences, whose ideas clearly go together. If he added a topic sentence and made the three sentences following more coherent and more emphatic, he would have a good paragraph.

Then suppose he took out the long quotation completely and kept only his *own* writing. He would then have these two logical paragraphs (remember that we are looking again at the *original* development):

```
     During the no-regulations period mining companies disposed
of their wastes in the most economical way. Often the refuse was
actually dumped in conical piles to encourage burning for shrink-
age of the area needed.
     At those times burning refuse piles were considered a neces-
sary, ugly, fuming evil whose extinguishment was almost impos-
sible or too expensive. People resigned themselves to the of-
fensive sight and smell of the gob piles.
```

Except for the illogic of "At those times" (it should be "at *that* time" to agree with "no-regulations period"), the last paragraph of the two has unity. But it is too thin: it does not have enough supporting material. Both "almost impossible" and "too expensive" need to be followed by specific details and examples (if examples can be found). To some extent the quotation serves as supporting material, but the points made are really generalizations. We would prefer to see specifics.

The first of these two paragraphs has unity, too, but there is only one detail (the second sentence) to give meaning to "in the most economical way." This paragraph, we see, is also too thin. Such short structures disturb the reader because they arouse his curiosity and then leave him "flat." His wants are not satisfied.

Paragraphing is always a challenge. As a composition in miniature, the paragraph must be just as logical and effective as the complete letter, report, memo, etc. However, some flexibility is not only permitted but also *encouraged.* For example, although report paragraphs generally should be fully developed, the writer should consider writing a short one—perhaps a single sentence—where it is appropriate. A short paragraph is especially appropriate when he wishes to emphasize a point, say, or to indicate that he is turning from one major idea to another.

Short paragraphs are even expected in such forms as letters and instruction manuals. In these forms they frequently contribute to both the emphasis and the empathy of the message. For example, many letters begin with a short, friendly greeting such as "Thank you for your letter asking for information about our line of laboratory equipment." Then the writer stops, indents for a new paragraph, and goes on with the next point that he wishes to develop. In instruction manuals, the short paragraph helps an operator of a machine or a maintenance worker follow directions easily.

Indeed, in all forms the use of a list (instructions in a manual, questions in a letter, conclusions in a report) often satisfies a reader's wants best. For his convenience, in fact, many writers indent all items in the list and then number the items in sequence. In a letter, for instance, when questions are presented in a numbered list, all the reader has to do in his reply is answer by referring to the numbers.

Perhaps the most important point about forms and the techniques used in them is that, when writing is good, they will not stand out. When a reader's eye is drawn to any form, he may very well miss something in the message. To be sure, the stereotype is not always the right form, and the writer then must create his own. But if the writer keeps his reader's wants in mind and then plans carefully, the reader will realize that the situation calls for an exception. If the development prepares him for a new form, he will *expect* (and accept) the new as much as he would the old in a routine situation.

EXERCISES

OVERALL FORMS

1. Do you as an individual now have opportunities to use overall forms for the messages you write? In each case, (1) describe the opportunity, (2) name the form you would use, and (3) explain why you would use the particular form you have named.
2. To convey a message as an individual, when could you justify a different form (telephone call, telegram, face-to-face meeting, gestures only) for communicating with someone to whom you would ordinarily write a letter?
3. Have you ever used a "wrong" overall form for a message? What was the situation in which you wrote the message? What form would have communicated more effectively in that situation? Why?

ORDERS OF DEVELOPMENT

4. For a particular message (class assignment) you have written, when have you made extensive use of one or more of the ten commonly used orders of development cited on pages 29–30? What was the nature of that particular message (assignment)? How extensively did you use the order(s)? Was your use of the form(s) effective? Why or why not?
5. Does it occur to you now that you could have used extensively one or more of the ten orders of development cited on pages 29–30, for a writing assignment that you completed recently? What was that assignment? What orders of development could you have used?
6. What order(s) of development can you identify in the following paragraphs?

    a. Although the infrared technique is relatively new, its applications to hydrology are numerous. Various forms of water pollution, especially thermal pollution, are easy to detect by infrared photography. The Canadians are interested in the results of infrared imagery which indicate current patterns in Lakes Erie and Ontario, and the influx and dispersion of thermal, industrial, and sewage pollutants. Infrared imagery has also been used to delineate tidal currents of polluted rivers over commercial shellfish beds in Merrimack Estuary, Massachusetts. In addition, infrared can be used to outline for mapping purposes many other features including drainage patterns, overgrown shorelines, and vegetation-covered jungle streams.

b. Liquid water in the form of clouds can have two opposite effects. During the overcast day, clouds reflect back into space radiation that would have reached the surface on a clear day. This process is a cooling one, and thus cloudy days are generally cooler than sunny days. At night, however, there is no incoming solar radiation. Clouds absorb and radiate back downward the long-wave heat radiation from the earth. This process tends to warm the surface, and cloudy nights are usually warmer than clear nights.

OUTLINES AS HEADINGS

7. In a fairly long piece of writing (500 words or more) you have done, examine your development to see if it is logical. You can check easily by composing topic heads now, if you did not then start with an outline. Could these topic heads have been included in the final draft of your writing? Would your writing have been more effective if the heads had been included?

8. How logical are the following outlines? Be specific in presenting your criticism.

a. Report title: "Manufacture and Use of Slurry Explosives"

```
I. Definition
   A. History
      1. Beginning
   B. Development
   C. Manufacturing Process
      1. Materials
      2. Machinery
         a. Blending
         b. Packaging

II. Use
   A. Places
      1. Ways in which slurry explosives are used
```

b. Report title: "The Microwave Oven"

```
1.0 Introduction
2.0 Theory of Microwaves
    2.1 Definition
3.0 Source of Microwaves
4.0 Oven Parts
    4.1 Theory of operation
        4.1.1 Cooking times
        4.1.2 Clean-up time
5.0 Commercial and Industrial Uses
    5.1 Effects on the population
```

PARAGRAPHS

9. To what extent does the paragraphing in a and b, following, lack unity, coherence, and emphasis? Do the paragraphs have—or appear to have—unity, coherence, and emphasis? That is, in each case, does the paragraph develop one major idea, develop it logically, and give prominence to key ideas? Is one idea fully developed? Are the transitions effective? Is the importance of ideas clearly indicated? Be specific in your criticism.

   a. Excerpt from a report entitled "Solving the Problems of Water Pollution in the Susquehanna River Basin":

   > The main problem here is not lack of treatment but inadequacy of treatment. Most cities and towns in the basin have sewage-treatment plants, and this is good, but the effluent is then dumped into the streams and here lies the problem. This treated effluent is harmful because it depletes the nutrients in the stream. The best technique to combat this problem is to institute waste-water projects. The effluent can be pumped for many miles from the treatment plant to a designated area. It can be sprayed on the land at the rates of from one to two inches per week. The University began a waste-water spray project several years ago and already has had excellent results. The results show that the effluent can be renovated by the complex of soil and plants so that 90-95 per cent of the detergents are held in the upper six inches of soil. Phosphorous was reduced by 99 per cent, and nitrates by 68-82 per cent. About 80 per cent of the water applied was recharged to the underground reservoir in a completely clean state. Irrigation can be conducted all winter. No evidence was found of nutrients in private wells of the tested region. The yield of crops increased 17-300 per cent, depending on the year and the crop. Conifers and hardwoods grew significantly in height and diameter in areas that were irrigated. A monitoring system has been devised to insure that the water reaches the ground reservoir in a clean state. During the winter months the nutrients are absorbed by the soil for use by next year's plants. No adverse effects have been found on wildlife--birds, mammals, or insects. The cost of these programs is small. The University has found that it costs only 18 cents per 1,000 gallons for a 10-million-gallon-per-day sewage plant (average for a town of 7,000).

   b. Excerpt from a report entitled "Extracting Sulfur Dioxide from Industrial Flue Gas":

   > It has been estimated that the average time sulfur dioxide remains in the air is less than 12 hours. During this time, only one-fifth is brought down in moisture droplets. The rest dissolves in water on buildings, in the soil, and in plants.
   >
   > The yearly cycle of suspended pollutants is much more

definite than that of deposited matter such as soot. At most places, the average concentration of smoke and sulfur dioxide in the winter is two to three times that in the summer. This is partly because of the extra fuel burned in the winter and partly because atmospheric conditions are more favorable in the summer for the dispersal of smoke and gases away from street level into the upper air.

Moreover, the toxic limit for humans is 500 ppm [parts per million] of sulfur dioxide during an eight-hour exposure. The limit is much lower in cold, damp weather, especially for people with bronchial or asthmatic problems.

The toxic limit for plants is 50 ppm. If this concentration is prolonged or exceeded, the plant loses its healthy green color, and the underside of the leaf is discolored.

During the London Smog of December 1952, the daily sulfur dioxide concentration reached 134 ppm.

Although sulfur dioxide is a colorless gas, it is affected by the ultraviolet rays of sunlight to oxidize to sulfur trioxide, which combines with water to form sulfuric acid.

The sulfuric acid in the atmosphere will then settle on buildings, bridges, pavements, etc., and lead to corrosion. It was estimated that 6 million dollars damage is done yearly by this type of corrosion in New York City alone.

The maximum ground-level concentration of sulfur dioxide will occur less than a half mile from the source of emission.

SPECIAL PROBLEMS

10. In a brief report (150–250 words) identify and discuss the effectiveness of the orders of development in the paragraph quoted in Exercise 10a.
11. In a brief report (200–300 words) identify and discuss the effectiveness of the orders of development in the paragraphs quoted in Exercise 10b.
12. Outline the ideas in the paragraph quoted in Exercise 9a. If you think the development is illogical, rearrange the outline.
13. Outline the ideas in the paragraphs quoted in Exercise 9b. If you think the development is illogical, rearrange the outline.
14. Rewrite the paragraph quoted in Exercise 9a.
15. Rewrite the paragraphs quoted in Exercise 9b.

Chapter Four

OBSTACLES TO UNDERSTANDING THE LANGUAGE

Finally in our analysis of the communication process, we come to language, the means by which communication takes place. Language is what the writer must encode and the reader must decode. At least in theory, we should note, language ultimately is the message in its most irreducible parts—word and syllable (individual sounds).

Of course, language is not *a* word. Language is *words*; and to have meaning words must be related to other words, and then combined and arranged with them. Since messages are made up of words, we shall begin our study by examining structures for relating, combining, and arranging words. We shall call these structures sentences, clauses, and phrases. We consider types of sentences and mood, voice, and order. Because they are the basic elements of sentences, clauses will be included in these analyses. Phrases, the smallest structures usually, will be examined after these. Then we shall look at language as words and style.

Use of the word "structures" may make you ask why sentences, clauses, and phrases are not included in Chapter 3, on forms. The reason is simply that these structures are groupings of words. Essentially, language covers what is *inside* the sentence, and form covers what is *outside*. Language covers the lesser structures; form covers the greater structures.

Note that this chapter refers to the handbook, which begins on page 279. The obstacles reviewed there are no less important than those presented here; indeed, pages 279-327 should be considered an extension of this chapter. At this point, however, we are concerned more with weaknesses in writing than with specific errors. This chapter covers *style*, the way in which a technical writer should combine, and even select, words. The handbook focuses on *errors in mechanics*, including spelling, punctuation, etc.

The final topic covered in this chapter is "readability" formulas, devices for determining how easy to read—and in many cases how clear—a writer's style is.

(You can check on your own style, too!) This section should help you review what you have read.

SENTENCES, CLAUSES, AND PHRASES

A *sentence* is a group of related words which stands alone because a complete thought is expressed or implied. It always ends with a period, a question mark, or an exclamation point; and it always contains, or implies, a subject and a verb. Three examples follow:[1]

```
The size of nuclear reactors varies greatly.
```
    (The subject is "size"; the verb is "varies.")

```
Is corn silage in itself an effective feed for lactating
cows?
```
    (The subject is "corn silage"; the verb is "is.")

```
No.
```
    (The subject and the verb are implied in this answer to the preceding question. "No" is an abbreviated statement for "Corn silage is not in itself an effective feed for lactating cows.")

A *clause* is a group of related words which may or may not stand alone. There are two types of clauses, and both contain a subject and a verb. A *main* clause expresses a complete thought and may either stand alone (that is, be a sentence), or be part of a sentence. A *subordinate* clause (a clause lower in rank) does not express a complete thought and never stands alone—as the name suggests. It is always only part of a sentence. The following example is a brief illustration.

```
The lens and the grating system are the most important
parts of the Spectronic 20 because they channel, or focus,
the light.
```
    (The main clause in this sentence is "The lens and the grating system are the most important parts of the Spectronic 20." This clause could also stand alone as a sentence because it expresses a complete thought. The subordinate clause is "because they channel, or focus, the light." This clause could not stand alone, because it does not express a complete thought.)

A *phrase* is a group of related words that does not contain a subject and a verb. Like a subordinate clause, however, a phrase always appears as a part of a sentence. The following sentence includes three phrases:

```
Begin the experiment by filling a flask with two ounces of
the solution.
```

---

[1] In technical writing, exclamations (sentences ending with !) are rare.

(The first phrase in this sentence is "by filling a flask"; the second is "with two ounces"; and the third is "of the solution." None of these three phrases contains a subject or a verb[2] or expresses a complete thought. Therefore, they cannot stand alone.)

Let us now examine these structures more closely to understand how they are used and abused.

TYPES OF SENTENCES

Based on the clause structure, a sentence can be classified as one of four types: simple, compound, complex, and compound-complex. Each type can be used effectively—but each can create a major obstacle to the reader.

Simple Sentence

A *simple* sentence is one main clause and nothing more. It has one subject and one verb, both of which may consist of several parts. It also may have one or more phrases. But it has only one clause. All of the following are simple sentences; subject and verb are underlined in each:

The pedicle flap also can be used as supportive tissue.

Both the Hayden and the Sharp methods were used.

Very little literature has been published on the use of sawdust or wood chips as a source of bulk for ruminants on high-energy rations.

The simple sentence is often used to emphasize a simple idea, as the illustrations above show. In the first example a distinctive, secondary use is noted. In the second, the writer is emphasizing the use of two particular methods. In the third, the emphasis is on the fact that not much has been written on the subject. These examples indicate that the simple sentence can be used to great advantage in our writing.

The simple sentence *can* be an obstacle to the reader's understanding, however. As you can see in the following excerpt, writing that consists of simple sentences only is difficult to read.

Thermal pollution is a type of water contamination. The contamination results from the continual discharge of industrial heat into rivers or lakes. The discharge raises the average temperature of the water about 15°. This rise in temperature may seem small. It can cause aquatic life to change, however. The increase is permanent. . . .

---

[2] "Filling" is a verb form used as a noun; it is not the verb of the sentence. See "gerund" on page 302 for a more complete explanation.

Since each grouping of words is given the same amount of emphasis, no grouping appears to have any more importance than another. Hence the reader must determine where the emphasis should be, and obviously he will have to read carefully. He may even have to *re*read. In either case his wants are clearly not satisfied.

How much more meaningful—and more comfortable to read—is this arrangement of the ideas:

```
    Thermal pollution is a type of water contamination resulting
from the continual discharge of industrial heat into rivers or
lakes. The discharge raises the average temperature of the water
about 15°. This rise in temperature may seem small, but it can
cause aquatic life to change because it is permanent. . . .
```

In the revision, the first sentence is effective because it completely defines "thermal pollution." For emphasis, the second sentence is best presented as a simple sentence, as it is in the original version. The third, and last, sentence combines the three ideas effectively because they logically go together. Indeed, because of the interrelationships, one might say that the clauses are inseparable.

Compound Sentence

A compound sentence is a combination of two or more main clauses. It is, therefore, an arrangement of simple sentences joined together. (Like the simple sentence, the compound sentence has no subordinate clauses.) All of the following examples are compound sentences. Again, the subject and the verb in each clause are underlined. A slash (/) is added to show where the clauses are joined.

The <u>machine</u> <u>is</u> then <u>switched</u> on,/and the transmittance <u>dial</u> <u>is turned</u> to 0°.

The <u>gap</u> <u>should be</u> small for efficiency,/but <u>it</u> <u>must be</u> large enough to ensure clearance.

Coordinating Conjunctions

The words following each slash in the above examples are called *coordinating conjunctions.* Coordinating conjunctions are used to join clauses that are equal in importance. In the first sentence, "and" joins two statements describing successive steps in an operation. Both ideas are equal; neither is more important than the other. In the second sentence, "but" joins two ideas which together establish a contrast. Again, neither idea is more important than the other.

Five coordinating conjunctions are recognized: *and, but, for, nor,* and *or.* (Some authorities recognize another, *yet.*) From the examples above, we can see that these connectors serve an important purpose in our writing. Often the technical writer has to report facts, conclusions, procedures, etc., that are equal in importance.

Limits on Use

Yet the very number of coordinating conjunctions limits their use in writing, generally. And probably this is why the compound sentence is used least often in report writing. You can readily visualize a report in which every structure was an arrangement of clauses joined by one of these connecting words. It is almost as bad to use only compound sentences as to use simple sentence after simple sentence. Some relationships and emphases are shown clearly, of course, but not enough—as we shall see next.

Complex Sentence

A complex sentence is a combination of a main clause and one or more subordinate clauses. It is used, obviously, when ideas to be expressed are unequal in importance. In the illustrations that follow, subject and verb in each clause are underlined as before, and again a slash is used to separate the clauses:

```
Since the water is enclosed in a pipe,/no water is lost by
evaporation.

The leveling-head assembly contains four screws/which are used
for leveling the transit.

For the purpose of this study, a holdover fire is defined as a
fire/that has burned at least eight hours/before it is dis-
covered.
```

Note that in the last sentence the second and third clauses are both subordinate clauses.

Subordinating Conjunctions

Complex sentences are common in practical writing because a great deal of work experience is concerned with causes, results, conditions, concessions, etc. For example, because something happens, something else happens—this very sentence should indicate how frequently you will be using the complex sentence. To emphasize the point, we examine a list of words, called *subordinating conjunctions*, that are used to introduce subordinate clauses:

To express cause: because, since, as, inasmuch as, whereas

To express comparison: as, as if, as though

To express condition: if, unless, whether . . . or, in case, so long as, provided

To express concession: although, though, even though, even if

To indicate place: where, wherever

To indicate purpose: that, so that, in order that

To indicate time: when, whenever, before, after, while, since, as, as soon as, until, till

As you can readily see, many subordinating conjunctions are available for your

use—and this list is not complete. Among other subordinating words that c
added are the relative pronouns—who, which, and that—which also are useu
introduce subordinate clauses.

Illogical Structures

As the result of faulty thinking, both compound and complex sentences are often
used erroneously. When a compound sentence is well developed, we call the
achievement *proper coordination*—that is, correct use of a coordinating conjunction. When a complex sentence is well developed, we call the result *proper
subordination.* Conversely, when the structures are illogical, we call them improper coordination and improper subordination, respectively. When such errors
are made, the conjunctions (and hence the sentence types) should be changed.
The following examples clearly show the faulty thinking in the writers' minds;
note the underlined conjunctions.

Improper Coordination	*Proper Subordination*
The primary element of the linear induction motor is the moving part, or armature, <u>and</u> is made of metal capable of conducting an electric current.	The primary element of the linear induction motor is the moving part, or armature, <u>which</u> is made of metal capa-ble of conducting an electric current.
Improper Subordination	*Proper Coordination*
Gamma radiation creates the most serious hazard, <u>as</u> ex-tremely thick shields must be used to protect workers.	Gamma radiation creates the most serious hazard, <u>and</u> ex-tremely thick shields must be used to protect workers.

The reader will surely have difficulty sharing the writer's message when errors
are made like those in the preceding sentences on the left.

Before we leave the discussion of conjunctions, we must mention one other
kind of connective: a *conjunctive adverb.* Conjunctive adverbs function as both a
conjunction and an adverb[3] and include such words as *therefore, however, also,
moreover, nevertheless,* and *instead.* As connectives these words link groupings
of words as both coordinating and subordinating conjunctions do. They are useful in that they increase the number of possibilities open to a writer.

Like the other connectives, conjunctive adverbs are *pivotal* words; that is,
the writer turns his thought at the point he uses a connective, and therefore he
makes one grouping—or a part of a grouping—more emphatic. Slashes are used in
the following examples to illustrate the pivoting function:

The heart of the system,/ however /, is an activated carbon-
fluidized-bed absorber.

---

[3] An adverb modifies a verb, an adjective, or another adverb. See page 294.

```
The compounds cannot be used on dairy cattle,/ therefore /, because
they will contaminate the milk.
```

```
The particles in the water are not in suspension;/ rather /, they
are in a colloidal solution.
```

Since connectives can help a writer to get unity, coherence, and emphasis in his structures, they must be used with care. Obviously, they should be used only where they are needed. Too many connectives—especially of the same kind—should not be used. Excessive use of *any* kind of word lessens its effectiveness. Also, the *right* connective for linking groups of words must be chosen. The large number of connectives available may make the choice difficult; but the writer should not use a pivotal word unless he is sure it turns the thought as the thought should be turned.

Finally, but no less important, the writer should be sure that the connective is placed in the appropriate place. In the first sentence of the three preceding illustrations, the writer apparently wanted to emphasize "the heart of the system." Had he wanted to emphasize the whole sentence, he would have put "however" either at the very beginning or at the very end.

Punctuation for connectives has to be precise. As is indicated in the handbook on pages 315-23, the use of commas, semicolons, and other marks of punctuation with connectives cannot be taken lightly. The writer's entire message may be decoded incorrectly if an error in punctuation is made.

Compound-Complex Sentence

A compound-complex sentence is a combination of two or more main clauses and one or more subordinate clauses. It is used when a writer regards a number of ideas as being too closely related to be separated into different sentences.

Suppose, for example, that a writer has the following series of word groupings in mind as a one-sentence structure. (Underlines and slashes are inserted again, and here brackets are added to show that the first subordinate clause divides the second main clause.)

```
Ordinary sand can be used,/ but [if high pressure is needed for
the fracture,] the sand often is crushed,/ and it therefore does
a poor propping job.
```

Occasionally, when groupings seem to require such "togetherness," a compound-complex sentence is quite appropriate. However, novice writers are tempted to add more groupings, because they think other ideas are also inseparable. The sentence above, for example, might well have been written like this:

```
Ordinary sand can be used, but if high pressure is needed for the
fracture, the sand often is crushed, and it therefore does a poor
propping job, so to alleviate the problem other agents have been
developed, such as glass beads, which are commonly used now and
which are very effective.
```

Clearly, the overdone compound-complex sentence violates unity, coherence, and emphasis. Furthermore, although a reader might decode one such sentence without irritation or acute discomfort, he would find it frustrating if sentence after sentence were developed like the preceding illustration.

For these reasons, compound-complex sentences are not recommended for most practical writing. If you limit the number of clauses—using simple, compound, and complex sentences almost exclusively—your writing will benefit in two ways. Not only will your reader be able to decode easily, but you also will find the task of writing easier.

OTHER WAYS OF CLASSIFYING SENTENCES

Sentences are classified in three other ways: by the *mood* and *voice* of the verb and by *order.* Explanations and warnings are in order to prevent obstacles to the reader's understanding of your language.

Mood

Mood is the form of a verb that shows how the action or condition indicated by the verb is conceived. Three kinds of mood are expressed: *indicative*, to state a fact or to ask a question; *imperative*, to command or to make a request; and *subjunctive,* to state a suggestion, a condition contrary to fact, a wish, a doubt, a regret, a concession, or a supposition.

Following are ways in which mood is handled in technical writing:

Indicative:
The wires are connected to the terminals. (States a fact)
Are the wires connected to the terminals? (Asks a question)

Imperative:
Connect the wires to the terminals.   (States a command)
Please connect the wires to the terminals.   (States a request)

Subjunctive:
I suggest that the wires be connected to the terminals.   (States a suggestion)
If the wires had been connected, there would have been no danger. (States a condition contrary to fact)

The importance of using the right mood at the right time is illustrated well by two kinds of *wrong* uses:

A worker has three different moods in the headings for the main part of his report "Reducing Evaporation Losses in Atmospheric Storage Tanks":
What Are the Common Causes of Evaporation?
Note the Inadequacies of Floating-Roof Tanks
Fixed-Roof Tanks Should Be Used

(Not only the headings confused the reader of this report; the development—as incoherent as the heading system—confused him also.)

At the end of a report, a worker presents his recommendations thusly:

```
Install oxygen-determining devices on the recovery-furnace stacks.
Install pyrometers on the lime kilns.  . . .
```

(The reader stopped after the second command; he was the plant manager and was not used to taking orders from subordinates.)

You should be consistent in handling mood in any one structure (including the outline for that structure). To be consistent, the writer of the report on evaporation losses would have to do some thinking first—and then reflect that thinking in both his headings and the report. For example, a more logical heading system is

```
Common Causes of Evaporation
Advantages and Disadvantages of Floating-Roof Tanks
Advantages and Disadvantages of Fixed-Roof Tanks
Conclusions and Recommendations
```

You should be discreet in presenting recommendations to fellow workers. Individuals react negatively when they are *told* what to do. The writer of the recommendations would have made a more positive impression if he had used the subjunctive rather than the imperative mood:

```
It is suggested that we take these steps to eliminate the
pollution problem around the plant:
    1.Install oxygen-determining devices on the recovery-furnace
      stacks.
    2.Install pyrometers on the lime kilns.
    3.etc.
```

Note that the structure of each recommendation is no different. What makes the difference is the use of subjunctive mood in the sentence preceding the list.

Voice

Voice is the means by which a writer indicates how an action is, was, or will be taken. There are two voices: *active* and *passive*. Active voice makes the subject of the sentence the actor, or doer; passive voice shows that the subject is acted upon (usually through some kind of an agency).

These examples illustrate:

```
Mr. Harvey connected the wires to the terminals.
```
(Active voice: Mr. Harvey, the subject, does the acting.)
```
The wires were connected to the terminals by Mr. Harvey.
```
(Passive voice: "Wires" is the subject; they are acted upon by Mr. Harvey, the agent of the action.)

Passive voice is used extensively in technical writing because workers are trained to report what happens rather than who makes it happen. But use of the passive does create problems. As the last example shows, the passive construction is more awkward than the active, and usually it requires more words. Also, in using the passive voice writers almost always omit the agency, but readers often want to know who was responsible for what happened.

Finally, use of the passive structure sometimes results in a nonsense error called a dangler. A dangler is an adjective phrase that really has nothing to modify but is made to appear to modify a word grammatically. This error is covered thoroughly on pages 000–000, but one illustration here will help you to see how confusing a passive construction can make a sentence:

Hurrying to complete the work, the wires were connected to the terminal. ("Hurrying" is underlined because grammatically it seems to modify the subject, "wires." "Hurrying" is obviously a dangler, however, because *wires* do not hurry. It was Mr. Harvey, who is not mentioned, who did the hurrying!)

Use of the active voice indicates what is going on in your work—the actions (movements, changes, sequences, etc.) involved. Only this voice can faithfully present the action you are writing about. Thus you should consider using verbs (and modifying adverbs) like those in the following list. These words are called *lively language* because they name actions that the human body can take. They enable your reader to *visualize* what you are describing. Verbs can be used alone, it should be noted, but they can also be combined with adverbs to express your ideas precisely. Although the list is incomplete, you can readily see that hundreds of combinations are possible.

Verbs

beat	draw	hold	press	shake	strike
bend	drive	keep	pull	shut	take
blow	drop	lay	push	sink	tear
break	fall	lift	out	stand	throw
bring	get	lower	raise	stay	tie
carry	give	open	run	step	touch
catch	go	pick	send	stick	turn
close	hand	point	set	stretch	work
dodge					

Adverbs

about	around	down	over
across	aside	forward	through
after	away	in	together
ahead	back	off	under
along	behind	on	up
apart	below	out	upon

As you think about possible groupings in which you can use such words, note that you can write active sentences without writing from the first-person (I) point of view. Machines, apparatus, etc., have functions, and their parts have functions. Thus you can show by writing active structures that they "take actions," too.

Orders of Words and Groupings

Sentence orders (orders of words and groupings) are identified as *normal,*[4] *periodic,* and *inverted.* Normal order is most common because it is the order in which we think most of the time. The subject (actor) comes first, the verb (action) next, and the object (thing acted upon) last. For example, *Mr. Harvey connected the wires to the terminals.*

Notice that in reading a normal-order sentence, you can stop before the end of most structures. Even the short example illustrates: you can stop after reading "wires" and find a complete, meaningful sentence to that point.

In periodic order, conversely, the first word or grouping is what would be last in normal order. Beginning and ending are interchanged. For example, *The wires were connected to the terminals by Mr. Harvey.* Periodic order is used to emphasize the idea at the end; and you must read to the very end before you will have as much information as you would if normal order were used.

In inverted order we see the third possible arrangement of the three basic sentence elements (subject, verb, object). Here the verb comes first. For example, *Did Mr. Harvey connect the wires to the terminals?* As this example illustrates, inverted order is used for asking questions. Also, although much less often, inverted order is used to gain emphasis by changing the position of words and groupings from what we expect. For example, *To the terminals Mr. Harvey connected the wires.*

A knowledge of orders is important to the technical writer because he then realizes how much the placing of words and groupings can affect the meaning. In normal order, no part of the sentence is given special emphasis: all ideas are about equal in importance (except that main clauses are always more important than subordinate clauses, of course). In periodic order, the last part of the sentence is emphasized. In inverted order, a word or a grouping that ordinarily is not stressed is given emphasis.

You can readily envision the obstacles that you will create by handling word orders thoughtlessly. It is apparent that normal order should dominate your writing; if you use too many periodic and inverted structures, emphasis is sure to be misplaced. Keeping the reader's wants in mind at all times is the only way you can be confident that the order you have selected for a sentence is "right."

---

[4]Called "loose" by some authorities; not so named here, to keep you from confusing writing that lacks unity, coherence, and emphasis—which is also called "loose"—with the desirable quality of writing indicated here.

PHRASES

Since phrases have neither subject nor verb, they must perform a special function in the sentence. Actually they perform two functions: they serve to establish relationships between basic structural elements[5] like the subject and the verb, and they give substance to these. They are links in the sentence and they give meaning to the sentence.

Clauses have little meaning without phrases. Consider the following sentence:

```
The time required for sampling with a soil tube depends on the
condition of the soil.
```

The words that are underlined in this sentence are phrases. If they are taken out, all that remains is

```
The time depends.
```

These three words may be considered a sentence but they share no information. Without the phrases, the "sentence" is meaningless.

Because phrases are so important, they too must be handled with care. In establishing relationships, they must be placed in the order that the thought of the sentence requires. In the sentence quoted, that order would be violated—and the meaning would change—if the first two phrases were interchanged:

```
The time with a soil tube required for sampling. . . .
```

Now the sentence seems to say, first, "the time *spent with*," and, second, "the *tube* is required." However, the relationships in the original sentence are clear: "required" modifies "time," and "with a soil tube" modifies "sampling."

Because phrases are so vital to your writing, you can see how necessary it is to arrange them correctly. If you can recognize the errors just illustrated—called *misplaced modifiers*[6]—in your own writing, you will begin to see that understanding relationships is the first requirement for developing effective structures.

Another problem, which may be even more serious, is the use of a wrong phrase structure for a context (grouping so closely connected to a word or words as to influence their meaning). Following are two examples of inappropriate phrases for the context:

```
The configuration of these heat exchangers is not in accord-
ance with ours.
```

```
Recommendations are based on the experience of other foun-
dries in relationship to this problem.
```

---

[5] Phrases can function as any element of the sentence—nouns, verbs, adjectives, etc.
[6] See also pp. 293-94, 310.

What each writer meant to say was, respectively:

```
The configuration of these heat exchangers is not like ours.

Recommendations are based on the experience of other foun-
dries with this problem.
```

As the original sentences show, the phrases themselves are not wrong; they are only wrong for the context. Such vague connective phrases would be perfectly all right in the sentence where they were appropriate to the writer's thought. However, this use often results in nothing more than wordiness—noise for the decoder. Other phrases that are frequently used in the wrong context are *as related to, as to, in connection with, in conjunction with, in regard to, with reference to,* and *with respect to.*

Good idiom and appropriate usage are earmarks of a careful reader and listener. If you don't have a good ear for them now, being more attentive may be all you need to develop one.

AWKWARDNESS

Awkwardness is an obstacle that is difficult to classify because it can be found in all structures. Awkwardness, simply stated, is *clumsy writing.* It creates a difficulty for the reader because the words in a grouping are not those he is accustomed to finding in such a structure. Awkwardness may reflect poor listening and speaking habits developed over many years.

Awkwardness can be found in both short and long sentences. In either, however, the clumsy handling of a phrase is almost always responsible for the weakness.

An example of a short and a long sentence follows; notice the underlined phrases:

```
The weakness of ether often requires that a greater concen-
tration of the drug be used.

When clouds are seeded, any noticeable effects brought about,
whether related or not, bring much criticism and even lead
to the banning of any other experiments in many areas.
```

In each of these sentences, the first underlined phrase is awkward. Note in each of the revisions, below, that subordinate clauses effectively replace these phrases:

```
Because ether is weak, a greater concentration of the drug
often must be used.

If the weather is at all unpleasant after clouds are seeded
in the area, people are critical whether or not the change
can be attributed to the seeding. Their criticism may even
lead to the banning of any other experiments in many areas.
```

These revisions show exactly what each writer *intended* to say, simply and directly. If you can spot phrases like the first in the original sentences, when you review your own writing, you may find that subordinate clauses like those in the revisions will effectively say what *you* intend.

Of course, other elements of the sentence can be awkward, as the second underlined phrase in the second original sentence illustrates. This phrase is part of the predicate (a verb-and-modifier, or "action," structure). The phrase is awkward on two counts: in itself (Did you ever hear of anything "to bring much criticism"?) and in the repetition of the verb "bring," which appeared earlier in the modifier form "brought."

Clearly, to write "gracefully," you need to use the structure that fits your idea exactly. Study and a continuing effort to write simply and directly are sure to help a writer structure sentences that are both less awkward and clearer.

THE WORDS OF LANGUAGE

To write simply and directly, begin by using words that are simple and direct. Naturally, the words must say what you *mean*. Remember, however, that the reader must be able to decode—to understand. Therefore, you must know as much as you can about ways in which you can help readers understand the information you want to share.

The first way to help readers is to realize that what you mean may be different from what *they* think you mean. Consider the individual word itself, for example. Do the words we all use in common—such as *country, Republican Party, governor, city council, plant manager, home, father, school, teacher*—mean the same, precisely, to all of you who are reading these words?

Words have denotative (dictionary) meanings and connotative (personal) meanings. The dictionary definition of home is "a house, apartment, or other dwelling serving as the abode of a person, family, or household."[7] However, to one of you, home may be synonymous with "love." To another, the connotation may be "security," to a third, "the establishment"—and so on.

At school or work, we tend to use words in their denotative rather than connotative meaning. Our personal words are our own; our dictionary words are everybody's. That is, our dictionary words *should be* everybody's. But even when we use the same words in what we think is the same way as others use them, we find at least some differences.

Take the word *impedance*. To those of you who are studying electricity, impedance means one thing. To those of you who are studying physics, it means another. Even among you electricity students, there may be a difference. For example, some of you may think *resistance* and *impedance* are synonymous; but as others of you know, the words are not interchangeable.

---

[7]Funk & Wagnalls *Standard College Dictionary*, New York: Harcourt Brace Jovanovich, Inc., 1963, p. 640.

A number of words in engineering and science are often confused. Following is just a partial list:

Word	*Word(s) often confused with*	
acceleration	velocity	
actual barometric pressure	sea-level barometric pressure	
adsorption	absorption	
compression	tension	
evaporation	vaporization	
formula weight	molecular weight	normal weight
frequency	period	oscillation
frost deposit	rime deposit	
interference	clearance	
jig	fixture	
kinematic friction	static friction	
mean	median	mode
normal time	standard time	
pressure	force	
reagent grade	technical grade	
tolerance	allowance	
vibration	damping	

Why are these words confused? Primarily, they are confused because those who use them are not always precise. Precision in the use of language, however, is essential if decoding is to be the same as encoding. To become acutely aware of the need for precision, you need to be exact in your use of everyday words—everyday, standard English.

Everyday English—Defined

What is everyday, standard English? Defined simply, it is the language any reader will understand because it is used—or could be used—in careful everyday speech. Note the phrases "any reader" and "careful everyday speech." They emphasize the point that if our language is to be clear at all, we must exclude the extensive use of

1. Polysyllabic (many-syllable) words derived from Latin and Greek
2. The jargon peculiar to specialized fields
3. Slang and colloquialisms (informal words)
4. Gobbledygook (the involved and repetitious use of words)

Simple Synonyms—for the Most Part

Even when you address superiors and fellow workers, write as simply and directly as you can. Prefer nouns like "change" and "forecast" to "emendation" and "prognostication"; verbs like "annoy" and "explain" to "exacerbate" and "elucidate"; adjectives like "heavy" and "juicy" to "ponderous" and "succulent"; adverbs like "slowly" and "wisely" to "dilatorily" and "sagaciously."

Here are other examples:

why say . . . ?	*when it is more natural to say . . . ?*
accumulate	gather
additional	added
ameliorate	improve
assistance	help
encounter	meet
endeavor	try
equivalent	equal
finalize	complete
initiate	begin
locality	place
modification	change
optimum	best
procure	get
terminate	end
utilize	use

Some Necessary Exceptions

Obviously, you cannot use one- and two-syllable words only, and some words derived from Latin and Greek are "everyday English." Thus, most of your readers will know them. Too, some words *must* be used, such as *antiseptic, brucellosis, cumulonimbus, deuterium, electrodynamics, flocculation, granulation, heliocentric, immigration, jurisprudence, kilogrammeter, leucopoiesis,* and *molybdenum*—to name just a few.

We must often use some words derived from Latin and Greek that fellow workers will expect. For example, if you were an engineer writing to another engineer about a symmetrical arrangement of parts in a camshaft, you would have to use the proper word—*configuration.* No other word says the same thing.

Some Unnecessary Exceptions

The following notes may help you to see when you should and should not make exceptions to the use of simple synonyms.

Some people believe that if you use words of Latin or Greek origin frequently, you make your message more concise. Conciseness—saying much in few words—*is* a virtue in practical writing. But the use of long words for the sake of conciseness is a *false economy.* To use frequently the long words in our language—such as "compensate" (make amends), "expedite" (speed up), and "facilitate" (make easy)—is to create difficulties for the reader.

Examine this example of false economy in writing:

```
Assuming thermometric and hygrometric unalterability, variation in
wrinkle resistance experimentation results attributable to treat-
ment is improbable.
```

The writer would need more words to simplify this sentence. But can't you read the following revision faster—and with greater understanding?

```
If the temperature and the humidity are constant during experi-
ments, treatment of the fabrics probably will not affect their
resistance to wrinkling.
```

The revision is not as concise as the original, but no one has to read the second version twice to understand it.

Another unnecessary exception to the use of simple words is the *specialist's abuse of technical language.* Most of the writing done by specialists is more technical than it has to be.

Consider, for example, a chemist's report on the advantages of a particular fabric dye. Of course, the chemist knows what "tinctorial process" means. But suppose the chemist's report is copied word for word in a letter to a client whose training was nontechnical? Wouldn't it be wise for the chemist, in planning his message, to think about writing so that any intelligent reader could understand? Here, for example, would it be unprofessional to write "dye process"? Or, couldn't he use both words to show that they meant the same thing?

Definitions are always needed in technical writing, even for the reader who has had technical training. A common complaint of managers and supervisors in industry is that they "don't know what the writer is talking about." "Sure, I graduated as a chemical engineer," an executive said recently while waving a junior chemical engineer's report. "But that was twenty years ago. With my regular duties, I don't have time to learn about every new development, every new theory, and every newly coined technical term in the field."

Equally disturbing to leaders is the perversion of standard English that tries to make the everyday word technical. The engineer who writes, "Follow this procedure to obtain the most satisfactory horsepower repeatability," when he could say, "Follow this procedure to keep horsepower at the same high level in each test" is making such a perversion. He is not making English work for him; he is working to use English. And his reader is going to have to work even harder to understand exactly what he means. By being a miser with his words and a spendthrift in his thinking, the writer is creating a major obstacle.

Excessive Use of Slang and Colloquialism

Some slang and colloquial expressions have become accepted as standard, every-day English through frequent and widespread use. Examples are *spot* in "Our company is in a tight spot"; *fired* in "The engineering department fired its chief metallurgist"; and *go* in "The design should go well with that of our other products." But overuse of such expressions, as in the following excerpt from a report, disturbs the reader:

```
As reported before, the sad results super-bugged the customer,
who already was pretty hyper about the lousy work our lab was
doing.
```

Like the writer, the reader probably thinks in language like this at times. He may even talk this way when speaking informally to workers. But he knows that informal writing like this for fellow workers may also appear in letters to customers. (No outsider would buy from a company whose writers used language as they pleased.) Hence, a leader will not let his workers have one set of standards for intracompany writing and another, different set for writing to people outside of the company.

Gobbledygook

Gobbledygook is just as easy to spot as excessive slang and colloquialisms. Gobbledygook is words, phrases, and clauses that are involved, complex, and long. Often, a very simple thought is hidden in the gobbledygook message, as this example shows:

```
May your annual commemoration of the winter solstice be abundant
with innumerable manifestations of felicity, and may the next cir-
cuit of the ecliptic engender a propensity toward bestowing
plenteousness and unparalleled prosperity.
```

Something like this message appeared on a greeting card a few years ago. Perhaps you were able to decode it as you read, for its thought, essentially, is "Merry Christmas and a prosperous New Year." Possibly, to impress on you the points on language made here, it's appropriate to end on a humorous note. But much of the gobbledygook in practical writing is anything but easy to decode and, therefore, anything but humorous.

READABILITY FORMULAS

Readability formulas are devices for determining how easy to read—and frequently how clear—a writer's style is. Although many such formulas have been developed over the years, only two will be reviewed here. Most formulas are created for the same purpose: to check any tendency you may have to write complexly. Specifically, they are designed to help you see if your sentences are too long and your words generally have too many syllables.

Gunning's Fog Index[8]

The "Fog Index" formula developed by Robert Gunning, an internationally known consultant on writing, has an interesting name. It is meant to be taken literally: the formula enables you to determine how "foggy"—that is, how "cloudy" or "murky"—your writing is. The answer you get in using the formula is the school-grade level at which your writing can be read.

To find this number, you take the following steps:

1. Count off in a sample of your writing to the period[9] closest to 100 words. Divide the total number of words in this passage by the number of sentences, which you must count, of course.
2. Going back over the same passage, count the words having three syllables or more, except
   (a) words that are capitalized (proper nouns).
   (b) words that are combinations of short easy words (like "bookkeeper" and "butterfly").
   (c) verb forms that are made three syllables by the addition of -ed or -es (like "created" or "trespasses").
3. Add the numbers representing the average number of words per sentence, and the number of words having three syllables or more. Then, to determine the Fog Index, multiply the sum of these two by .4.

Written as an equation, the formula is

$$.4 \: [\text{ASL (average sentence length) + PS (percentage of polysyllabic words]}$$
$$= \text{Fog Index}$$

The school-grade reading level and the appropriate Fog Index value for each are indicated in the following chart. To give the chart more meaning, Gunning shows a magazine having the same Fog Index:

	Fog Index	Reading Level by Grade	Reading Level by Magazine
	17	College graduate	
	16	" senior	(No popular magazine
	15	" junior	this difficult.)
	14	" sophomore	
Danger Line	13	" freshman	
	12	High-school senior	*Atlantic Monthly*
	11	" junior	*Harper's*

---

[8] From *The Technique of Clear Writing*, revised edition, by Robert Gunning, New York: McGraw-Hill Book Company, 1968, pp. 38–40. Used by permission.
[9] Or a semicolon (;) or colon (:) dividing two main clauses.

	Fog Index	Reading Level by Grade	Reading Level by Magazine
	10	High-school sophomore	*Time*
Easy-reading	9	" freshman	*Reader's Digest*
Range	8	Eighth grade	*Ladies' Home Journal*
	7	Seventh "	*True Confessions*
	6	Sixth "	Comics

Damerst's Clear Index

The formula developed by the author is similar to the Fog Index in the bases used. However, instead of producing a value of grade-level reading ability, the Clear Index shows a number value corresponding to grades given in schools. By using this formula, a writer will find a value, between 0 and 100, which represents the "grade" he earns if his writing is very difficult to read, on the one hand, and very easy, on the other.

Note that in using this formula you adjust the total number of words in a sample, so that you always divide into 100 to obtain the average number of words per sentence. You also adjust the number of words, proportionately, when you count individual words. Finally, note that instead of counting words of three syllables or more, you count only those words that have one syllable—that is, one sound.

The series of steps for working out the Clear Index follows. Once you try the formula, you will find it easier to use than the detailed instructions seem to indicate.

To find the Clear Index of a piece of writing:

1. Count off about 100 words, stopping at the period (or the semicolon or the colon separating main clauses) that is closest to 100. Record the exact number of words, counting everything that is not an actual word (for example, $25) as one word.

2. If the number is higher or lower than 100, adjust it to 100 by changing the number of words above or below 100 to a percentage. (Example: the total to the closest period is 115; therefore, change the 15 over 100 to 15%.) Then apply the percentage to the total. Although precision is important later—in steps 4 and 5—it is not essential here. (Thus in our example we can obtain 100 approximately by subtracting 15.)

3. Reread the same piece of writing; as you read, count the number of words having only one syllable (sound). Count contracted words (like "we'd") as single words. Record the number. (In our example let's say we find 67 words having one sound.)

4. Adjust the number of one-sound words by using the same percentage that was applied in step 2. (Example: the number found in step 3 was 67. Thus

the adjusted number of one-syllable words is approximately 57 (67 – 15%, or 10; 67 – 10 = 57).)

5. In the scale below, on the "Number of One-Syllable Words" line, locate the number found in step 4. Directly below the number you will find the "Clear Word-Value." If the number of one-syllable words lies between two numbers on the top line, add or subtract 2.5 to the "Clear-Word Value." This number should be recorded in a separate place. (In our example the "Clear Word-Value" is 25.)

The Clear Index Word-Value Scale											
Number of One-Syllable Words	47	49	51	53	55	57	59	61	63	65	67 or more
Clear-Word Value	0	5	10	15	20	25	30	35	40	45	50

6. Find the average number of words per sentence by dividing the number of sentences into the number found in step 1. Remember to count main clauses as sentences if they are separated by a semicolon or a colon. On your work sheet, record *this* number; do *not* adjust it. (In our example let's say that we find six sentences. The average number of words per sentence, then, is approximately 19 (115 ÷ 6 = approximately 19).) You may be exact if you wish, but in most cases the approximate number is close enough.

7. In the scale following, on the "Average Number of Words per Sentence" line, locate the number found in step 6. Directly below the number you will find the "Clear Sentence-Value." Add this value to the value found in step 5. The sum is the Clear Index. (In our example, the Clear Sentence-Value is 40. The Clear Index of the writing in the example, therefore, is 65 (25 + 40 = 65).)

The Clear Index Sentence-Value Scale																					
Average Number of Words per Sentence	27	26	25	24	23	22	21	20	19	18	17	16	15	14	13	12	11	10	9	8	7
Clear Sentence Value	0	5	10	15	20	25	30	35	40	45	50	50	50	50	45	45	40	35	25	10	0

To show that using the formula is less difficult than may appear, let's apply it to the first six sentences of the chapter (through "and arranging words"). The analysis is presented by the numbers of the seven steps:

1. This sample contains 102 words.

2. A count of 100 is found by changing the "2" above 100 to 2% and deducting 2%—or 2—from 102.
3. The sample has 70 words of only one syllable (sound).
4. The 70 words become 68 (68.6 actually) when we apply 2% and subtract 2 (1.4 actually) from 102.
5. According to the first scale, the "Clear Word-Value" is 50.
6. Since sentence 2 of paragraph 2 is actually two sentences, separated by the semicolon after "Language is *words*," we divide 7 (sentences) into the 102 total. By division we thus find that the average number of words per sentence is 15 (14.57 actually).
7. According to the second scale, the "Clear Sentence-Value" is 50. The Clear Index of the writing in this sample, therefore, is 100 (50 + 50).

What does each of these indexes mean? In both, it is apparent that the ideal number of words per sentence is 14 to 17—on the average, of course. Even more significant, though, is the complexity of the language. A large number of multisyllable words is sure to make reading difficult—to create a major obstacle to the reader's understanding. If you use the Fog Index formula, the number of words of three syllables or more should not, on the average, be greater than 6.

The Clear Index formula is even more restricting. It indicates that a high percentage of words of more than one syllable will create an obstacle. The premise of this index is that most of the words we use should have only one syllable. Indeed, in a 100-word sample, there should be at least 67 one-syllable words for a Word Value of 50 to be used. After all, you will remember, it is the *simple*, one-syllable words that are easiest to understand. These words come to us mostly from Anglo-Saxon roots—the roots of the everyday English language we use today.

Note that when the average number of words per sentence drops below 14, the Clear Sentence-Value drops also. The reason for this is that extremely short sentences are likely to make one's style jerky. And the shorter the sentences are, the more difficult it is for the reader to see how they are related. In the extremely short sentence, there is little room for transitional (linking) clauses and phrases.

Before we leave indexes, several points should be made. First, if you use an index, use it for what it is. It is only a check on writing to see if there is too much fog or too little clarity. It will not teach you how to write, of course, and therefore may seem to be of questionable value. But as a check, to tell you if your sentences are increasing in length or your language is getting more complex, an index can be useful. It should indicate how much difficulty your reader will probably have in decoding your message.

Secondly, take several samples of a single piece of writing—not just one. Our style of writing—our language, if you will—changes as we touch on different topics. The more technical we must become, the more complex and difficult our language seems to become. But in his book *The Technique of Clear Writing*,

Robert Gunning has a chapter on technical writing that shows how the Fog Index can be kept down to 10 or so.

And if you stop to think of it, by writing simply and directly—as recommended throughout this chapter and in the handbook—a low Fog Index or a high Clear Index should be possible. When you define, when you include explanation (for example, by analogy and by other techniques for simplifying), you cannot help using more roots-of-our-language words than the other, more complex ones. You'll see for yourself if you try.

EXERCISES

SENTENCES, CLAUSES, AND PHRASES

1. Identify the clauses and phrases in the introduction to this chapter (first five paragraphs).
2. Identify the clauses and phrases in your last writing assignment.

TYPES OF SENTENCES

3. Identify the types of sentences in the introduction to this chapter (first five paragraphs).
4. Identify the types of sentences in your last writing assignment.
5. Identify the coordinating conjunctions, subordinating conjunctions, and conjunctive adverbs in the introduction to this chapter (first five paragraphs).
6. Identify the coordinating conjunctions, subordinating conjunctions, and conjunctive adverbs in your last writing assignment.
7. Determine if the coordination or subordination in the following sentences is logical or illogical. Improve each sentence that you think can be made more logical—and therefore more effective.
   a. These data were requested by Mr. Huntly and were taken from tests run on a production engine.
   b. Since I have read your proposal for improving the system, I agree that the system could be better than it is.
   c. This test unit is the least expensive of all, and by using it we will be able to reduce the total cost of the project.
   d. Some of the employees are not doing their job, and several projects have not been completed on time.
   e. Often our chronotach breaks down in the middle of a test, and we lose a half day and cannot adhere to the supervisor's schedule.

OTHER WAYS OF CLASSIFYING SENTENCES

8. Identify the moods in the following sentences.
   a. Was the experiment finished on time?

    b. The value of the work now being done in the Research Department is questionable.

    c. We recommend that the operation be discontinued at once.

    d. Discontinue the operation at once.

    e. The supervisor asked the worker why the report presented had not been completed.

9. Identify the voices of the verbs in the sentences in Exercise 8.

10. Identify the sentence orders in the sentences in Exercise 8.

PHRASES AND AWKWARDNESS

11. Identify and correct the errors in idiom in the following sentences.

    a. Most of the workers at the plant are aware and concerned about the amount of pollution created by the plant's effluent.

    b. It is amazing for me how much more new employees know than those who have worked here for years.

    c. The supervisor had a definite recommendation with reference to solving the problem.

    d. From his book, the author presented many of the new developments of which I had been unacquainted.

    e. His action was exactly in accordance with the action he took in this situation last year.

12. Identify and correct the awkwardness in the following sentences.

    a. One success of the research to find new foods is the process for manufacturing proteins from paraffins.

    b. I have read several recent articles in your magazine that have been thorough in the description and solution of the problem we have at our plant.

    c. All coverage that you requested has been dealt with in the accompanying report.

    d. Printed matter in the form of leaflets used by this union in its organizational efforts is foremost in quality among all major unions.

    e. Tapping emergency water supplies and implementing curtailment regulations have been helpful in ensuring minimum gallonage during the drought.

THE WORDS OF LANGUAGE

13. Look up in your dictionary the meanings of *chemistry, physics, engineering, nuclear,* and *microbiology.* Are the meanings you find the same meanings that you have always assigned to these words? Discuss.

14. List some words that you are familiar with in your field of study. Then look up the words in your dictionary. Do they have the same meaning in the dictionary as the meaning you have always assigned to them?

15. In your opinion, can each of the following words be replaced by one or more simpler words? a. accommodation, b. condominium, c. generalization,

d. multitudinous, e. recrudescence. Discuss your answer in each case.

16. What does each of the following sentences mean? (*Hint:* each is a wordy statement for a well-known simple saying.) (The sentences are taken from five issues of *Effective Letters*, copyright, New York Life Insurance Company.)

   a. It has come to our attention that herbage, when observed in that section of enclosed ground being the property of an individual other than oneself, is ever of a more verdant hue.

   b. A warm-blooded vertebrate of the class *avis* grasped in the terminal prehensile portion of the upper limb of the human body is equal in value to one plus one of the aforementioned vertebrates in a shrub.

   c. It has been observed that an enclosing barrier, for the purpose of discouraging and preventing intrusion upon that which it encloses, tends to enhance the amicability of those whose property abuts on said barrier.

   d. A mineral matter of various composition when engaged in periodical revolutions exhibits no tendency to accumulate any of the cryptogramic plants of the class *musci.*

   e. It has been observed that the individual who devotes himself continuously to gainful pursuits, and thus neglects the cultivation to be derived from diversion, emerges thereby as a listless and uninteresting personage.

READABILITY FORMULAS

17. Using Robert Gunning's Formula (page 60), find the Fog Index for the writing in paragraphs 3 and 4, page 42, in this chapter. (Of course, stop counting as close to 100 words as possible.)

18. Using *either* the Fog Index *or* the Clear Index formula, determine how readable (a) your last writing assignment was and (b) your present writing assignment is. Take at least several samples in each case and write a brief report listing the values found and comparing the readability of your writing in each assignment.

SPECIAL PROBLEMS

19. In revising your last writing assignment, attach a report in which you note specifically the changes you have made that were influenced by your reading of this chapter.

20. Using *either* the Fog Index *or* the Clear Index formula, determine how readable the writing is in at least five textbooks you are now using (or have used) in other courses. Take at least three scattered samples in each case and write a report listing the values found and comparing the readability of the authors' writing. Discuss the role that the complexity of the subject matter has *before* you rate the readability of each book. (Be sure to ask yourself, however, if a textbook containing complex subject matter is written as simply as it could be.)

MASTERING THE SKILLS
OF TECHNICAL WRITING

Chapter Five

RESEARCH AND INTERPRETATION

The basis of communication is research. *Broadly defined, research is the seeking of information.*

The extent to which research is carried out depends on what is to be learned—fact only, fact and interpretation, or interpretation only. Since fact and interpretation are key terms, we need to understand both clearly.

A *fact* is a truth that is known by observation or experiment, or by some other means of verification. It is anything that can be proved to be some *one* thing always. For example,

The density of oxygen gas is 1.43 grams per liter.[1]

A fact, by definition, cannot be ambiguous (have more than one meaning). It does not say that a thing has relative value, for example, that it is good or bad, high or low, sharp or dull, etc. Indeed, a fact means the same thing to everyone; it is an absolute.

An *interpretation* is a meaning given to two or more related facts. A fact by itself means nothing—except that it is a fact. An interpretation may *seem* to be possible for the statement "The density of oxygen gas is 1.43 grams per liter." You might say that this density is high, or that it is low, and claim that you have an interpretation. But if you think about "high" and "low," you will realize that either interpretation can be made *only* if you have other information. Both a highest density value and a lowest density value (in grams) and two values (in grams) in between had to be determined. Only when three ranges—"high," "medium," and "low"—had been established as factual constants, and the ranges established by deduction, could the claimed interpretation be made.

An interpretation such as "This density is high" or "This density is low" is helpful when a researcher needs no more information. Usually, however, he needs more. The two related facts below, followed by an interpretation made for a practical reason—an important decision—illustrate:

---

[1] It is assumed that a constant, realistic pressure applies.

Fact: The density of oxygen gas is 1.43 grams per liter.

Fact: The density of liquid oxygen is 1.20 grams per milliliter.[2]

Interpretation: For the space ship's air supply, liquid oxygen is much more practical than oxygen gas because the liquid form requires less storage room.

Whereas a fact means only one thing, an interpretation can have several meanings. An interpretation can be an *evaluation*, a *conclusion*, an *hypothesis*, or an *idea.* We must be sure that we understand these terms, too.

THE MEANINGS OF INTERPRETATION

An *evaluation* is a determination of either worth or amount. The writer of the preceding interpretation made such a determination. When an evaluation is to be made, two or more things, theories, processes, etc., are compared for the purpose of naming one as the better (of two) or the best (of three or more). Of course, a standard always exists. In the air-supply example above, compactness was the criterion. In another case, the criterion might be expense, say, or availability.

A *conclusion* is a judgment or opinion based on the relationship found between two or more facts. A conclusion may lead to an evaluation. For example, the evaluation of the two densities (liquid oxygen vs. oxygen gas) was based on a conclusion. The conclusion, of course, was that liquid oxygen was more compact.

An *hypothesis* is an unproved scientific conclusion. Like a conclusion it is drawn from known facts; unlike a conclusion it is not an end but a beginning—a basis for further experimentation or investigation. When an evaluation or conclusion is stated, the related facts needed in a research study have been collected —and meaning has been found in them. When an hypothesis is stated, the facts needed for the study have not yet been collected. When they are, and they *prove* the hypothesis, the hypothesis can be restated as a conclusion or evaluation.

For example, the researcher of the facts in the earlier illustration would have started with the following hypothesis:

For the space ship's air supply, liquid oxygen is much more practical than oxygen gas because the liquid form requires less storage room.

At that time he would not have known both densities. To prove the hypothesis, he would have to get the exact information about the densities. (If the information were not available, he would have to experiment to get it.)

Naturally, if the research is completed but the hypothesis is not proved, a different conclusion must be stated. Or, if the early results of research point to an-

---

[2]In this experiment, it is assumed that a constant, realistic pressure applies for both oxygen gas and liquid oxygen.

other hypothesis, perhaps a change can—and should—be made during the research. We shall look at such possibilities again in the section "Using Imagination During the Research Procedure."

An *idea* is a thought resulting either from the mind's working consciously or from the exercise of imagination. It differs from a fact in that it has not yet been verified. It differs from an evaluation or a conclusion in that it does not have the finality of a determination of either worth or amount, or of a judgment or opinion. Of course, an idea may lead to an evaluation or a conclusion; but first an hypothesis must be stated and research must be carried out.

Usually, an idea is a wondering or a supposing. The following expressions illustrate:

I wonder what liquid oxygen's density is.

Is the density of liquid oxygen high or low?

Suppose the density of oxygen gas is too low.

What happens if we compare the densities of oxygen gas and liquid oxygen?

Maybe liquid oxygen would be better because it has a higher density.

Isn't liquid oxygen better because it has a higher density than oxygen gas?

RESEARCH

Research is classified as pure and applied. *Pure research* is a seeking of information without a defined goal. An example is undirected laboratory experimentation, which is carried out for its own sake. *Applied research* is a seeking of information for a practical reason—for a use to be made of the results obtained. An example is experimentation to increase the efficiency of a detergent. Applied research is always undertaken to attain a goal.

Of course, pure research often results in a practical discovery; and applied research may stimulate one to do pure research. Too, a scientific discovery may be made accidentally. (The process of "cracking" in oil refining is but one example of a valuable discovery that was made by accident.)

Whether research is pure or applied, all of the sources of information available to man can be used. Sources of information are classified as *secondary* or *primary*. Let's look more closely at each classification.

Secondary Sources

Secondary, or secondhand, sources give us information gathered and recorded by others. Examples are books, encyclopedias, dictionaries, reports, pamphlets, brochures, journal articles, magazine articles, newspapers, and company records. From these sources we gain such information as

1. Facts and general information.
2. Results of experimentation, along with details of procedure, apparatus used, methods followed, etc.

3. Statistics presented in tables, charts, and graphs.
4. Computations, diagrams, and drawings, along with photographs.
5. Interpretations—sometimes with recommendations.
6. Theories and ideas for further work on the subject.

Secondary sources are always a good place to start research. The results, interpretations, and recommendations of others may be vital to the success of our own work. Indeed, we may have to look no further and may not have to do any work of our own. We never want to unnecessarily repeat work that another researcher has done.

Work on any project may well begin with what is called a "search of the literature." This search takes us first to the *indexes* (alphabetical lists) and *abstracts* (short summaries) of books, articles, reports, etc. In a particular study, we may want to refer to one or more of the following sources; the list is not complete, of course, but it covers many fields:

Aeronautical Engineering Review
Aerospace Engineering Index
Agricultural Index
Applied Mechanics Reviews
Applied Science and Technology Index
Bibliography of Scientific and Technical Reports
Biological Abstracts
Business Periodicals Index
Chemical Abstracts
Dairy Science Abstracts
Electronics and Communications Abstracts
Engineering Index
Excerpta Medica
Field Crops Abstracts
Forestry Abstracts
Geoscience Abstracts
Horticultural Abstracts
Industrial Arts Index, The
International Aerospace Abstracts
International Index to Periodicals
Journal of the Institute of Petroleum Abstracts
Meteorological and Geoastrophysical Abstracts
Mineralogical Abstracts
New York Times Index, The
Nuclear Science Abstracts
Readers' Guide to Periodical Literature
U. S. Government Research Reports

In addition to these guides to sources and information in abbreviated form are almanacs, dictionaries, encyclopedias (of biological sciences, chemical technology, physics, etc.), guides, handbooks, and yearbooks.

Primary Sources

Primary sources originate with the worker. They are the sources by which he gets information firsthand, as a result of his personal effort. Experimenting and observing and asking questions of others (through face-to-face or telephone interviews, printed questionnaires, and letters of inquiry) are the means of getting primary data.

Experimenting and Observing

When we think of experimenting and observing, we usually think of a laboratory. The work of such specialists as chemists, physicists, and their assistants is readily identified with, for example, distillation columns and electric circuits. Using their knowledge, these people work under controlled conditions to perform such tasks as analyzing compounds and transferring energy. The laboratory is the only place where such work can be accomplished and where precise observation can be ensured.

However, we should note that specialists may have to work in the field, as well. Actual tests under real conditions are more meaningful than simulations ("like-real" experiments) in the laboratory. For example, the laboratory is the place to synthesize new compounds for tire materials. But highways are needed to test the finished tire on cars and trucks traveling at a variety of speeds for thousands of miles. Similarly, the highway pavements under those tires must be tested after development in the laboratory. Field work often is needed to prove that what happens in the laboratory, happens in the real world, too.

Of course, observation involves all five senses—hearing, touching, smelling, tasting, and seeing. Eliminating the ugly gray billows from a paper mill's smokestack is a problem, to be sure. But a greater problem is getting rid of the rotten-egg smell of hydrogen sulfide (and of other stack gases) permeating the surrounding air. Frequently, more than one sense is needed to confirm the successful completion of an experiment.

Asking Questions

When we think of asking questions of others, we usually think of a face-to-face meeting—a personal interview. But sometimes we can do better by using the telephone, mailing a questionnaire with a return envelope, or sending a letter of inquiry. Certainly, by asking questions through one of these other means, we can contact many more people.

However, the advantages of the personal interview are obvious. By seeing the reaction of the listener we know when he or she does not understand our questions. We can also discount answers when the listener's expressions and actions indicate that he is being evasive.

We can reach more people by telephoning, but we can't be sure that we have called at a favorable time. If we haven't, our listener may answer quickly but glibly—just to get rid of us.

There are disadvantages in using the mail questionnaire and the letter of inquiry as well. However, if the reader is sincere and interested, he most likely will take time to answer, and to answer thoughtfully. He doesn't have to answer the moment the questionnaire or letter of inquiry arrives.

Phrasing Questions Precisely

All questions should be composed with great care. If all that you want is a "yes" or "no" answer, or an "either-or" answer, you should word the question so that no other answer is possible. Such a question is called a *two-way question.* For example,

```
Are the liquid and the gas fed into the column at the same
time?
                                          Yes ____    No ____
```

The advantage of the two-way question is that it is easy to answer. It is, therefore, a good "ice-breaker": those answering are likely to be more interested in your survey if you begin with simple questions.

The two-way question is weak, however. Often—as in its use in the example shown—it must be followed at once by another, or *follow-up* question. For example,

```
If your answer to the preceding question was "No," should
the liquid be fed into the column before or after the gas?

              Before the gas ____    After the gas ____
```

When you use follow-up questions, you increase the length of the interview, questionnaire, or inquiry. The longer the list of questions, the less interested the listener or reader becomes. Two-way questions should therefore be used sparingly, preferably at the beginning.

More meaningful, usually, is a question affording three or more possible answers. Not only will you be able to ask fewer questions, but you will avoid forcing a "yes" or "no" answer when the listener or reader would prefer to give neither. In such a case, he would like to check a "Don't know" or "Not sure" answer.

The question that provides three or more possible answers is called a *multiple-choice question.* Following is a simple example:

```
Is the phenol separated from the chloroform during extrac-
tion, during distillation, or at some other time?

During extraction ____  During distillation ____  At some
other time ____
```

This illustration covers all answers. At times, you will have to end the list of possible answers short of complete coverage. When you do, you should end the list with an answer such as "Other (please specify)." By adding "please specify" in parentheses, you actually provide for several other answers.

To obtain the greatest amount of information, you should use what is called an *open-end* question. This kind of question invites the listener or reader to present several sentences in reply. For example,

```
To what extent have you used natural control of the insect
population in your area?
```

Notice here how the writer has changed a rather meaningless two-way question ("Have you used . . . in your area?") into one that forces the listener or reader to reply thoughtfully.

Since the object of asking questions is to get the greatest amount of information you can, you should be able to make good use of multiple-choice and open-end questions in your survey work.

Pretesting and Sampling

Pretesting is a means of finding out in advance how effective a list of questions is in obtaining useful answers. A few listeners or readers are asked the questions, and their answers are analyzed. If the questions are clear, and the answers look promising, the full-scale survey can then be carried out. If the questions apparently are not clear, the pretest should show what corrections should be made before the greater survey is begun.

You can test the technique by trying out your questions on fellow students in your field or some friends, regardless of their fields. Either should be able to tell you if your questions are clear, at least.

Sampling is a means of covering a large group by asking questions of only a few. In sampling, one tries to pick the few so carefully that he obtains from them everything that he wants from the group.

There are two ways to pick the few. One is called *representative* sampling; the other, *weighted* sampling. Representative sampling, as the name suggests, is used to cover the whole of a population to be surveyed. No member of that population can be excluded from possible inclusion in the few. Weighted sampling is used when a select (that is, limited) group of a whole population is to be surveyed. Representative sampling might be used to find out what brands of personal soap Americans use. Weighted sampling might be used to find out what brands of detergents housewives use.

Although it is not always realistic to sample a bibliography of secondary sources, the technique *can* be used in researching secondary sources. Suppose, for example, that you have a large number of books, articles, reports, etc., for a study on the general subject "transistors." By looking at abstracts and tables of contents, you might decide to read only a few sources because these covered the subject completely. In this case, you would decide that the other sources would only repeat what was said in the few selected.

As you know, sampling is common in the laboratory. You may not realize, however, that both representative sampling and weighted sampling are done. To illustrate, let us consider the assignment of testing a gallon of a solution. First, you pour the solution into test tubes. Then, to do representative sampling, you pick, say, any twelve of these for your testing. Thus you sample the whole gallon.

Now suppose you must also do some weighted sampling. In some of the test tubes a brownish-colored precipitate forms. You therefore test only the tubes having the brownish-colored precipitate—or only those whose precipitate is white, or yellow-brown, or some other color. When, as in this case, you test for one variable, you are actually working with a weighted sample.

Tabulating

When you experiment and observe for the same thing a number of times, or ask the same questions of a number of people, you accumulate a great deal of data. Your next step is to sort and record the data. *Tabulating*, or listing, is the way to make such a record. You simply count the values or the answers of each type, and indicate clearly what you have counted.

The process is easy, you will find, except when test results lie outside predetermined ranges, or when answers to open-end survey questions are all different. These data should not be discarded, but they must be examined carefully before a meaning—or no meaning—can be assigned to them. Sometimes a great deal of significance can be attached to data that cannot be tabulated.

NOTE TAKING AND DOCUMENTATION

Since a report, article, or other communication is the end product of research, you must take notes during your work with both secondary and primary sources. Then you must plan to identify all of your sources, with complete details so that the reader can go to the sources himself, if he wishes.

Note Taking

There are three kinds of notes for recording information other than that you obtain from your own experimentation and observation: quotation, paraphrase, and summary.

The *quotation note* is a word-for-word reproduction of what another person has either written or spoken. This definition points to two requirements: the writer's or speaker's words must be copied *exactly*, and quotation marks (" ... ") must be used[3] to set these words off from your own.

---

[3] In a report or article, when a quotation is three-or-more lines long, quotation marks are not used. Instead, the writer usually indents the quotation (uses a narrower margin that that for his own text) and single-spaces, rather than double-spaces, the quotation.

The *paraphrase note* is an indirect quotation. It is a restatement in your own words of what was said in a source by a writer or speaker. Thus written or spoken words may *not* be copied exactly, and quotation marks may *not* be used.[4]

The *summary note* is an abstract, or capsule statement, of lengthy writing or speaking. Obviously, summary notes do not include word-for-word copying, and quotation marks cannot be used.

The general word for calling attention to the ideas of others in your writing, is *citing*. Quotation, paraphrase, and summary notes are all citations. Citations are useful in finished reports, articles, etc.:

1. when you realize that you need to show your reader that you have used more than your own resources;
2. when the statement of an authority on the subject agrees with your own interpretation and therefore strengthens your evaluation, conclusion, or idea;
3. when a quotation says an idea better (more colorfully, more emphatically, or more gracefully) than your own words would say it.

You should therefore give serious consideration to the use of citations in your writing.

Cautions to Observe in Taking Notes

Although the ideas of others can be very helpful, your citations should be both brief and infrequent, for the most part. Rarely should you quote more than a phrase, a clause, or a complete sentence or two. Longer citations are necessary only when you must quote a law in full, say, or summarize another experimenter's procedures, results, and interpretations. In technical writing, citing another experimenter's work is always necessary when it provides background information for your own experimental work.

If you quote, paraphrase, and summarize others' words too often, you open yourself to the charge that the ideas in your report or article are not really your own. The impression you give is that you merely compiled others' ideas and filled in, with transitions, etc., between them.

A very important point about citing is that you must acknowledge the ideas of others so that the reader will not think they are your own. If he does think this, you are guilty of *plagiarism*—appropriating as your own the words and ideas of others. It will not matter to him that you did not mean to plagiarize. Plagiarism is plagiarism whether it is intentional or unintentional.

Thus, you must be extremely careful when you are taking notes. If you copy an original phrase or an original sentence word for word, remember to enclose in quotation marks everything that you copy. Also, be sure not to use quotation marks when you truly paraphrase or summarize, unless part of what you write is copied word for word.

---

[4]In a report or article, quotation notes may be included in both paraphrase and summary notes in the finished draft.

The Procedure for Note Taking

For recording both others' and your own information, you will find the following procedure helpful:

1. Use 3 × 5 (or 4 × 6) cards for notes. (An inexpensive alternative is to use quarters of 8½ × 11 sheets.)
2. Prepare a tentative outline. Then put on separate cards each topic and subtopic of your tentative outline. Attach index tabs of different colors so that topics, subtopics, etc., can be distinguished. Arrange the "index cards" in the order that your tentative outline places them.
3. Put only one note on each card. Then if you change your outline, you can move the cards to wherever the new outline indicates.
4. For each source, list the author's complete name (and, if possible, some indication of his authority for "speaking") and complete details for the bibliography, or "References" page, of your report or article.

The specimen card shown illustrates this technique.

Effect of DDT on Wildlife - birds
Paraphrase: The effects of DDT on the body chemistry of birds is strange. Somehow – the process is unknown – DDT checks the production of calcium, which is the major element in egg shells. George Laycock, "Pesticides Anyone?" _Field + Stream_, LXIII, Nov. 1968, p. 117.

To avoid repeating source information on every card, you can, if you wish, assign a different number or letter to each reference used. Then you need only make sure that you have complete source information for each reference.

5. Put only one question of an interview or questionnaire survey on each card and show the nature of the answers received, along with the number giving each answer. One card might well be used for the details about the date and time of the survey, the kind of sampling done, the size of the group answering, etc.

6. For your own experimental work and observations record the details of single tests, their results, and your interpretation(s).

7. Use separate cards for each idea you may get during research. Be sure to label each idea *as your own.*

If you are in doubt about recording any information, put it down on the appropriate card. It is better to have too much information than to trust your memory.

Documentation

Documentation is the supplying of details about a source. The common system of documentation consists of footnotes and a bibliography; footnotes appear on each page where citations appear, and the bibliography appears at the end of the report or article. Another system worth looking at has only brief documentation in the text itself, and a list of "references" at the end. We shall study both systems because each is used in technical writing.

Footnotes and Bibliography System

The Footnote The footnote, as the name suggests, is presented at the bottom of the text page. The first footnote on a page is separated from the text by double-spacing and a ten- to fifteen-space underscore. Each footnote is numbered to agree with the number following the citation in the text above. It is therefore easy for the reader to find the footnote when he drops his eye from the text to the bottom of the page.

Before we look at typical footnotes, let's review a list of abbreviations that appear in notes. This list is short, but it includes the abbreviations used most frequently. The italicized words and abbreviations are Latin; well-known abbreviations such as e.g., are no longer italicized by many authorities.

Abbreviation	*Meaning*
ch., chs.	chapter(s)
col., cols.	column(s)
ed., eds.	editor(s); edition
e.g. (*exempli gratia*)	for example

et al. (*et alii*)	and others
f., ff.	and the following page(s)
fig., figs.	figure(s)
ibid. (*ibidem*)	in the same place
i.e. (*id est*)	that is
loc. cit. (*loco citato*)	in the place cited
n., nn.	note(s)
n.d.	no date (of publication) shown
no., nos.	number(s)
op. cit. (*opere citato*)	in the work cited
p., pp.	page(s)
[sic] (*sic*)	thus (presented in brackets to show that what precedes is quoted exactly as it appeared in the original publication)
v. (*vide*)	see
vol., vols.	volume(s)

Although it is impossible to catalogue all kinds of footnotes, the following are typical. The examples are presented in the order and the form in which they would appear in a finished report.[5] Commonly used abbreviations are shown where they are needed. Explanatory comments are included in brackets, to guide you in your own use of footnotes. (Because the footnotes are for illustration only, they are fictitious.)

1 John Wills, Advanced Chemistry, 4th ed., New York: Hayes, Grimes and Company, 1970, p. 36. [Book]

2 "Space Age Chemistry," Springfield, Mass. Mirror, August 21, 1971, p. 18, col. 3. [Newspaper article]

3 Wills, op. cit., p. 44. [Same as Footnote 1 but different page]

4 Ibid., p. 46. Same as Footnote 3 but different page. Ibid. is used because Will's book has just been cited.]

5 Roger L. Murray, ed., New Horizons in Chemistry, Eaton, Calif.: Eaton Publishing Company, 1969, pp. 202–204. [Book edited by Murray.]

6 John Wills et al., Chemistry for Tomorrow, New York: Hayes, Grimes and Company, 1968, p. 53. [Book written by many authors. Wills's name appears first either because he contributed more than any other author or because he was responsible for the entire book.]

7 Murray, loc. cit. [Same as Footnote 5. Can you see why Ibid. must not be used here?]

8 V. Table 2, p. 3. [The reader is referred to a table which was presented and discussed on a previous page in this report.]

---

[5]If the list is actually presented like this at the end of a report, it is called, appropriately, "end notes" rather than footnotes.

9   Wills, Advanced Chemistry, p. 39. [Op. cit. cannot be used because two works by this author have been cited.]
10 Ibid. [Exactly the same as Footnote 9.]
11 G. B. Zirn and L. N. Harris, "Potable Water from Desalination," Water Resources Journal, February 16, 1971, pp. 99–103. [Magazine article by two authors.]
12 Dr. Joseph D. Allerton, Professor of Chemical Engineering, Eastern State University, in a personal interview, November 11, 1971.
13 Chicago Journal-Times, October 18, 1971, p. 40, col. 1. [Newspaper article without title.]
14 "Desalination," National Encyclopedia, 7th ed., vol. IV, p. 61. [Encyclopedia article with author's name not shown.]

As these footnotes show, when a book or an article bears the author's name, the author's name appears first. When the work is edited, the editor's name appears first. When the name of the author or editor is unknown, the title appears first. When the title of, say, an article is unknown, the name of the publication appears first.

As the eighth and twelfth examples show, at times you may have to compose your own form. It is expected that you will present the note in your own words when a conventional footnote form does not exist.

The Bibliography

The big difference between footnotes and the bibliography is the order. Footnotes are numbered, according to where the citation is made in the writer's text. The bibliography, on the other hand, is always presented in alphabetical order, by the first letter of the author's or editor's last name. If there is no author or editor, then the first letter of the title is used.

Other differences should also be noted. For books, the author's name and the title are followed by periods instead of commas. Other elements of the entry are separated by commas. For articles, a period appears after the author's name; from that point on, commas are used to separate elements to the end. As a convenience, some writers show inclusive page numbers (for example, pp. 47-52). In effect, these authors are saying, "The publication cited has more pages, but I made use only of those shown."

For an illustration, we shall look at the bibliography for the preceding list of fourteen footnotes.

Chicago Journal-Times, October 18, 1971.
"Desalination," National Encyclopedia, 7th ed., vol. IV, p. 61.
Murray, Roger L., ed. New Horizons in Chemistry. Eaton, Calif.: Eaton Publishing Company, 1969.
"Space Age Chemistry," Springfield, Mass. Mirror, August 21, 1971.
Wills, John. Advanced Chemistry, 4th ed. New York: Hayes, Grimes and Company, 1970.

```
Wills, John, et al. Chemistry for Tomorrow. New York: Hayes, Grimes
    and Company, 1968.
Zirn, G. B., and L. N. Harris. "Potable Water from Desalination,"
    Water Resources Journal, February 16, 1971.
```

Note that the interview (Footnote 12) is not included, because it is not a published work. Instead, it would be placed under the heading "Interview" (equal in degree to "Bibliography"), after the last entry in the bibliography. If answers to letters of inquiry were cited in the report, "Correspondence" would be handled in the same way.

Of course, the bibliography may include more books and articles than are listed in the footnotes. Books and articles sometimes are read only for background information; no citations are made. However, all of the sources cited in the footnotes must appear in the bibliography.

Anticipating the next section, we might note here that some simplification and abbreviation are becoming acceptable in the handling of the footnote-bibliography system just described. Some abbreviations, such as "Jour." or "J." for "Journal," are becoming popular among journal editors who continually look for ways to shorten articles. Others, such as the use of numbers only for the volume and issue numbers of a journal, are standard among a fairly large number of editors. In this modification of the system, the numbers are presented simply as "44:191"; the first number is the volume number, and the second is the issue number.

Nevertheless, the old footnote-bibliography system perseveres. You should therefore check pages 79-80 before you begin abbreviating freely. No matter how you handle documentation, however, you should be consistent: once you start using abbreviations, you will find it very easy to make careless errors.

Brief-Documentation and References System

The brief-documentation and list-of-reference system is used by many technical and scientific writers. The references are listed, not in alphabetical order, but in sequence as each work is cited for the first time. On the "References" page, each book and article is numbered in sequence, in the order cited. The citation itself appears in parentheses only as the number of the reference, followed by the page(s) cited.

For example, in the bibliography shown above, the last item (seventh in the list but the fifth new citation in the report) would be listed as follows in the "References":

```
5.  Zirn, G. B., and L. N. Harris. "Potable Water from Desalina-
       tion," Water Resources Journal, February 16, 1971.
```

In the *text itself*, where this article was cited, the entry would be:

(5, pp. 99-103)

The parenthetical note always comes immediately after the reference is made to the source's information. Many writers even drop the abbreviation for page(s) and write simply the reference number and the page number(s). In the example cited, the note would be simplified to (5, 99–103).

Although some experts believe that this system distracts the reader, it has two obvious advantages. It not only simplifies documentation but also eliminates the duplication in the footnotes and bibliography system.

LOGIC AND ILLOGIC

Logic

Scientific research calls for strict adherence to the scientific method. As you probably have learned in your training, the scientific method may begin with experimentation, with observation, or with the statement of an hypothesis. You discover something while experimenting or observing; or an idea comes into your mind. Then you work with this discovery or idea until a formal hypothesis takes shape in your thinking. Next you test the hypothesis many times. You either "prove" or "disprove" your hypothesis when you interpret the results obtained.

At times, of course, you must revise your hypothesis while you are still doing your research. Problems arise and things "just don't work out." With a new hypothesis, you may find that they will. Strict adherence to the scientific method requires an open mind throughout the research experience, whether you are in the laboratory, in the field, or in the library.

To have an open mind, you must use two logical-reasoning processes: induction and deduction. *Induction is the logical process of assembling facts until a conclusion is reached.* If the researcher identified at the beginning of this chapter were doing pure research only, his thinking would likely be as follows. He would observe that the density of oxygen gas was 1.43 grams per liter. Then he would observe that the density of liquid oxygen was 1.20 grams per milliliter. These two facts, when related, would lead him to the conclusion that liquid oxygen was much denser than oxygen gas. This is an *induction*—it is the basic method of thinking in science, the way in which natural laws are discovered.

However, induction is not the only way of thinking. Once the natural law is discovered and proved by experimentation, the scientist naturally wants to apply it where he can. Such an application in thinking is called a *deduction*, the use of a natural (general) law in a specific case. In the case cited, for example, our experimenter would be thinking deductively if he started with the conclusion that the denser a thing is, the more compact it is. His thinking would likely be as follows: things that have high densities are compact; liquid oxygen has a high density; therefore liquid oxygen is compact.

Of course, our experimenter would be thinking of the comparison of liquid oxygen and oxygen gas. He would be doing applied research to find a compact source of an air supply for a space ship. Hence, he would continue his deductive thinking, as follows. A source of air supply that is compact is desirable; liquid oxygen is compact; therefore liquid oxygen is desirable. Oxygen gas would not now enter his thoughts, because in comparison it could not be called compact.

Analogy is a deductive process. To say that the flow of electricity through a wire is like the flow of water through a hose is to deduce, by a knowledge of the characteristics, that the processes are similar. One must know both processes well, and recognize the similarity, in order to conclude that the analogy is appropriate. When he can make such a conclusion, probably he will realize that presenting the analogy in his report may be very helpful to his reader.[6]

Studying cause-to-effect and effect-to-cause relationships is an exercise in logical thinking, too. However, both inductive and deductive thinking can be used. If one knows the causes, he may by judicial thinking be able to induce a conclusion about the likely effects. Or, if he knows the causes and also knows that at another laboratory these causes produced a certain effect, he may be able to deduce that the same effect will apply in his case. If one knows the effects but not the causes, he may be able to conclude what the causes were. Or, if he knows the effects and also knows that at another laboratory these effects were brought about by certain causes, he may be able to deduce that the same causes can be found in his case. To be sure, investigations must be carried out to prove these inductions and deductions.

Finally, we should realize that logical reasoning is required even when only facts and factual generalizations are sought. Although in such a case the facts to be collected on a subject may be obvious, the researcher must be sure to include only the facts he needs and to leave out those that do not apply. The parts of a study must add up to the whole, and the scientist has to use interpretation to determine what the parts should be.

Thus, you can see, scientific thinking is likely to be a combination of induction and deduction. It goes without saying that the facts, causes, etc., must always be related and that the natural law or the generalization applied must be appropriate to the individual case.

Illogic

You should be continually aware that your thinking can be illogical as well as logical. If it is illogical and your reader sees that it is, he may begin to question everything you write. Therefore, during research you should be on your guard against falling into one or more of the following "illogic traps."

To practice applying your understanding of logical thinking, determine which process—induction or deduction—has been used (wrongly, of course) in the case cited in each illogic trap.

---

[6]It should be noted that although analogies simplify and clarify, they are never used to prove anything.

Reliance on Questionable Authority

Most sources in your research can be used without question. However, you always want to be sure that records cited really apply and are up-to-date. You should also be sure that authorities whose statements are used for support are qualified to speak on the subject of your *limited* study. It is one thing for a town or city official to voice his concern over the inability of the sewage plant to process an increase in wastes. It is quite another thing to use his comment to support your recommendation of a centrifugal pump of a *certain* size.

Hasty Generalization

Especially when you work in a laboratory, you should be careful not to become overenthusiastic when the results of your first few tests support your hypothesis. You should never jump to a conclusion just because all of the early evidence points one way. Many tests are considered necessary in the laboratory, and a number of observations—not just one or two—in field research.

Wishful Thinking

When you either interview or conduct a survey, you should be careful in classifying neutral, "don't know," and other uncommitted answers. You should not, for example, draw a conclusion like that made by one researcher. He did a study to find out the potential market for an electric drill having a radically different design. The following illogical conclusion and substantiation appeared in his report.

```
     The results indicate that we will have a large market for the
new drill. Of those surveyed, 26% said that they would buy the
drill if it performed as claimed and more than 50% expressed at
least some interest.
```

Sweeping Generalization

When you write, "All authorities say," and "Everyone filling out the questionnaire agreed," be sure that such statements are honest. You must be prepared to answer such questions as:

How many authorities did you consult?

Can you be so definite if you read only one or two articles and stopped because you found two authorities in agreement?

How many people, of the total surveyed, returned the questionnaire?

Even if, say, all 29% of those who replied gave the same answer, can you report that the *entire* sample gave that answer?

Those to whom you address your reports always raise questions when you give them *any* reason to doubt. Therefore, you should always qualify when categorical statements cannot be made. You should say, for example, "Four of the experts on fuel economy," and "Thirty-three percent of the chemists surveyed."

Substantial Majority

When you survey by mail or interview, you should remember that 51% or a slightly higher value does not indicate substantial agreement when only two answers (such as "yes" and "no") are given. In such a case, no one could use the phrase "substantial majority." The best way to handle low figures is to give the actual percentage. If it is 51%, you can honestly say only something like, "A very slight majority, 51%, ..."

Hedging

On the other hand, when the evidence is conclusive, you should not hedge. You should explicitly state the conclusion and provide supporting data. If, for example, a value found in a long series of tests is always the same or very nearly the same, you should say, strongly, "This value indicates . . ."—not, weakly, "It may be that this value indicates . . . ."

Misstatement of Problem

Problems neither identify nor state themselves. The burden falls on you, the investigator; and sometimes the burden is made heavier by a vague assignment such as "Please look into this." Yet the specific purpose and scope of the problem must be stated accurately and clearly if you are to find a solution. Note the following misstatements and the improvement made by the revision in each case:

Incoherence

Nonsense: How can we aid efficiency in the production department?

Improved: What can we do to speed up production? (This is what the writer meant. What would he have found, do you think, if he had stayed with the nonsense statement?)

Failure to state as a problem

No problem
indicated: The free silica in Benson iron ore concentrates.

Improved: How can the free silica be removed from Benson iron ore concentrates? (When a problem is stated as an open-end question, it is clear that a problem exists and that a solution will be sought.)

Failure to restrict the problem

Too broad: How can air pollution be controlled?

Improved: How can we meet the County limit of .5 lb of dust per 1000 lb of stack gas? (Had the writer stayed with the "too broad" statement, he could easily have written a book! Actually, he had "air pollution at our plant" in mind, but the specific problem to be solved at the time was the violation of the County ordinance.)

Begging the question

Question begging: What, if anything, should be done to treat the sulfite waste effluent at the Preston mill?

Improved: What can be done to lower the sulfite content in the waste effluent at the Preston mill? (The question that comes to mind in reply to the first question is, of course, "Should anything *be* done?" The original problem question was formulated without serious thought. The writer *knew* there was a problem but he begged the question just stated by taking for granted the very matter in dispute.)

SPECIAL USES OF INTERPRETATION DURING INVESTIGATIVE WORK

All research requires *some* interpretation. Even the strictly factual study places demands on the investigator; he must not only clearly identify his subject but also indicate his intention and show his coverage of the subject. Then, too, the investigator seeking only facts must determine what outline he should follow to present his material to the reader when his research is completed.

Naturally, when a research project calls for more than facts, the investigator must make greater use of interpretation. Special uses of interpretation that are frequently made are:
1. Solving a problem
2. Evaluating by standards
3. Justifying a conclusion
Let us look more closely at these three uses of interpretation.

Solving a Problem

Unless the researcher knows the exact cause(s) of a problem, he will have to use interpretation to analyze the problem (break it down into its parts). Consider, for example, the work of an engineer seeking the solution to the complex problem of underproduction in his company. He will proceed logically by investigating in detail the efficiency of the present machinery, production line, quality control, work supervisors, and work force. Those parts of this problem that he finds to be contributing to the underproduction are properly called subproblems.

After identifying each subproblem, he will try to obtain all the relevant facts about each and interpret them. When he has resolved all the subproblems, he will then use interpretation to develop a solution to the overall problem.

The logic of these steps is reflected in the structure of the report written after the investigation has been completed. The writer of a problem-solving report usually begins with a statement of the problem; then he lists the subproblems into which the overall problem divides; and next he analyzes each individual subproblem, in turn. The solution he presents at the end is based, of course, on the conclusions he has drawn in interpreting the facts about each subproblem.

The following brief excerpt from a problem-solving report illustrates the use of interpretation in this kind of study. Taken from a study of the contribution to air pollution in a community made by an unsupervised refuse bank, this excerpt covers the subproblem of excessive ventilation.

```
     Also responsible for the problem created by fires is the
amount of ventilation in the bank. Conspicuous from a distance
are old timbers (including wedges and cribbing pieces), rusty
lubricant barrels, brattice cloth, broken machinery, and large
rocks. However, as the close-up photograph (Figure 4, opposite),
together with the size analysis (Table 1, below), also shows,
much of the settled material is quite coarse (68% of all the
refuse is from 1 to 6 inches in diameter). Thus, there are many
voids in the bank, and oxygen flows through these freely. When a
fire starts, it always spreads rapidly.
```

Evaluating by Standards

When the researcher knows the exact cause(s) of a problem, his use of interpretation may be rather mechanical. All he may have to do is determine which of the solutions available is the better (of two) or the best (of three or more) possibilities. For example, when production machines are found to be inefficient, the researcher proceeds logically by evaluating two or more machines that can do the work required. Here, he may have to use interpretation only to select one machine over the other(s).

The process of evaluation requires the researcher to apply one or more standards. A standard is an authoritarian model or measure which is used to guide one to a decision on the worth or amount of something. To illustrate: for a particular experiment, a researcher finds that a distillation column made of carbon steel must be used. However, the column in his laboratory is made of stainless steel. Therefore, his decision can only be that his distillation column does not meet the standard and should not be used.

In technical work, a standard is called a specification when a specific material (as just illustrated), a precise dimension, etc., is called for. In some specifications a tolerance—that is, a small range of values—is permitted. For example, a tolerance of from 440 to 460 volts may be allowed for the voltage reading of a compressor's motor.

Obviously, when specifications and tolerances are stated, the investigator's use of interpretation will be very mechanical. But when standards are not so clear-cut, he may have to use interpretation a great deal. Indeed, he may have to relate a number of facts and draw several conclusions before he can create each standard for the new, radically different design of an apparatus, a machine, a process, etc.

Conventionally, the author of an evaluation by standards develops his report in one of two ways. One way is to name (and discuss, if necessary) *all* the standards first and then show how well the subject being evaluated meets each standard. The other way is to name *one* standard and then show how well the subject being evaluated meets that individual standard. After this, the next standard is presented and a comparison is made, and so forth. This development is often used when a number of standards, rather than just a few, are applied. After the evaluation is completed by either of these developments, the writer ends his report by stating (and discussing, if necessary) his overall evaluation. Naturally, if more than one subject is being evaluated, the author covers all the subjects after he presents the standard.

To illustrate these two ways of developing evaluations, we shall look first at excerpts from a report evaluating automobile mufflers. The excerpt that follows is from the list of standards, which the writer presents before the evaluation section:

A muffler's effectiveness in reducing the amount of pollutants can be measured against standards imposed by the State of California. This state has established two standards for the amount of pollutants: the maximum allowable concentration of unburnt hydrocarbons is 275 ppm (parts per million); the maximum allowable concentration of carbon monoxide emitted is 1.5% of the total emissions.

The second excerpt from this report is from the evaluation section following the presentation of the standards:

As far as reduction of the amount of pollutants is concerned, the Universal-Martin muffler was found to reduce the hydrocarbon content to 186 ppm and the emission of carbon monoxide to 0.62%. Both of these values are substantially lower than the standards.

Now we shall look at an excerpt from an evaluation report in which each individual standard is followed immediately by the comparison. The subject of the report is the evaluation of cellulose acetate as possible membrane material for the purification assemblies in the reverse-osmosis process of purifying water.

The first characteristic to be considered is the selectivity of the membrane. Selectivity is the ability of a material to separate salts from water. The membrane used must have a high degree of selectivity to achieve the salt-rejection factor required to prepare potable water from seawater. As was noted in the introduction, the salt-rejection factor for the purification of salt-water is approximately 100.

The second excerpt is the paragraph that immediately follows the paragraph just quoted:

> The cellulose acetate membrane is theoretically able to achieve the necessary salt-rejection factor for seawater, and to achieve it at relatively low values of driving-force, or pressure. In practice, however, these theoretical values are not easy to obtain. Imperfections inherent in the cellulose acetate material allow saltwater to leak through. In fact, therefore, it must be said that cellulose acetate does not meet this standard.

The first sentence in this paragraph is worthy of note. It is a concession included on the assumption that the reader may know what cellulose acetate is able to achieve *in theory.* Having made the concession, the writer can proceed convincingly to show that cellulose acetate fails on this one count. Obviously, we should observe in passing, for this reason the writer will have to judge this material to be unacceptable.

The third excerpt from this report—the overall evaluation—includes such a judgment. Note the amount of interpretation.

> Although the cellulose acetate membrane is an extremely good filter and thus requires very little pretreatment to reduce the concentration of suspended solids, its replacement cost is high, its permeability is poor, and, most important, its selectivity is offset by the imperfections inherent in the material. Indeed, cellulose acetate is unacceptable as a membrane material on the basis of the imperfections alone.

Justifying a Conclusion

Of course, all interpretations include justification; one must always have the facts in order to validate a conclusion. However, a special use of interpretation is to justify a conclusion against actual or anticipated opposition. The conclusion is strongly supported; the challenge to the writer is to convince the reader of its validity.

The writer of a justification also uses one of two developments conventionally. Either he begins with his conclusion and then justifies it, or he begins with his justification and then presents his conclusion.

To illustrate, we shall look at a justification for a large expenditure, which can be recovered only over a long period of time. The writer is an engineer employed by a firm which has just begun the manufacture of printed circuits by a dry-screen process. This part of his report is his support for justifying the use of a computerized drill to establish a standard for the size of the hole in the circuit board.

> Drilling the holes is a problem for all circuit-board manufacturers. Normal tolerances, of $\pm$ .010 inch from true center, are easily obtained with the use of a manually operated drill

press running at 3600 rpm. However, the manually operated drill
press is inadequate for drilling boards that require close tolerances, ± .002 inch from true center. Since a majority of customers' orders today specify close tolerances, a computerized
drill--which alone is accurate to ± .001 inch--is needed.

In this paragraph, the author presents his justification first, you will note; his
conclusion, last.

IMPORTANCE OF STANDARDS IN ALL INTERPRETATIONS

A keen awareness of standards is needed in all interpretive work. A researcher,
for example, will be aware of standards for equipment and apparatus, procedures,
chemicals used, observation of results, computations, preparation of graphs—
even for evaluation of the literature cited on the subject and for presentation of
footnotes and the bibliography.

During a research project, all aspects of a subject that require interpretation
can be reexamined in terms of standards. And, naturally, because the standards
exist, frequent comparison of what is being studied with the requirements means
that evaluation is continually taking place, also. The role that standards play in
any interpretive investigation is clearly very important to successful completion
of the study.

IMPORTANCE OF USING IMAGINATION
IN INTERPRETIVE WORK

In the final analysis, a scientist's and an engineer's thinking must be judicial. A
sound conclusion must be drawn whether it is made by induction, deduction, or
a combination of both. However, as was indicated early in this chapter, hypotheses and ideas are important, too. Indeed, if an idea is the prime mover in getting
research started, we need to examine its function in the technical writer's experience.

Ideas may come from conscious thinking, but equally as often they come
from the exercise of imagination. Many techniques and methods for solving
problems imaginatively have been worked out within the past 40 years. One of
these, called "brainstorming," was the invention of the late Alex F. Osborn.
Although the word may suggest harum-scarum thinking to some, Dr. Osborn
conceived brainstorming as a *disciplined procedure for discovering useful ideas.*
It is a group technique, basically, for from eight to twelve people, although an
individual, sitting alone with pencil and paper, can use it effectively, also.

The brainstorming group includes a leader "to keep the ideas coming" from
members, and a recorder to make sure that the ideas produced are put on paper.
The discipline is achieved by strict adherence to the four Osborn rules:

1. *Criticism is ruled out.* Adverse judgment of ideas must be withheld until later.
2. *Free-wheeling is welcomed.* The wilder the idea, the better; it is easier to tame down than to think up.
3. *Quantity is wanted.* The greater the number of ideas, the more the likelihood of winners.
4. *Combination and improvement are sought.* In addition to contributing ideas of their own, participants should suggest how ideas of others can be turned into *better* ideas; or how two or more ideas can be joined into still another idea.[7]

The basis for Rule 1 is that people will not suggest ideas if they are likely to be criticized. Rule 2 implies that *any* idea should be volunteered; an idea may be foolish, but it may suggest another that is practical. Rule 3 indicates that it is less desirable to pick from just a few ideas than to choose from 40 to 50 (a number a group can produce in less than ten minutes). And Rule 4 reminds us that the mind often works most effectively by making associations; brainstorming helps people to conceive relationships and associations.

Of course, ideas do not always come to us easily. For this reason, Osborn developed a list of what he called "idea-spurring" questions. His suggestion was that we ask these questions of the subject being brainstormed:

1. *Put to other uses?* New ways to use as is? Other uses if modified?
2. *Adapt?* What else is like this? What other idea does this suggest? Does past offer parallel? What could I copy? Whom could I emulate?
3. *Modify?* New twist? Change meaning, color, sound, odor, form, shape? Other changes?
4. *Magnify?* What to add? More time? Greater frequency? Stronger? Higher? Longer? Thicker? Extra value? Plus ingredient? Duplicate? Multiply? Exaggerate?
5. *Minify?* What to subtract? Smaller? Condensed? Miniature? Lower? Shorter? Lighter? Omit? Streamline? Split up? Understate?
6. *Substitute?* Who else instead? What else instead? Other ingredient? Other material? Other process? Other power? Other place? Other approach? Other tone of voice?
7. *Rearrange?* Interchange components? Other pattern? Other layout? Other sequence? Transpose cause and effect? Change pace? Change schedule?
8. *Reverse?* Transpose positive and negative? How about opposites? Turn it backward? Turn it upside down? Reverse roles? Change shoes? Turn tables? Turn other cheek?
9. *Combine?* How about a blend, an alloy, an assortment, an ensemble? Combine units? Combine purposes? Combine appeals? Combine ideas?[8]

A point that you must observe during your own experimentation is that you

---

[7] Alex F. Osborn, *Applied Imagination*, rev. ed., New York: Scribner's, 1957, p. 84.
[8] Osborn, *op cit.*, p. 318.

apply no question so that ill-effects occur. That is, you should not apply an idea if there is a possibility that property will be damaged or lives endangered.

The mind works judicially and creatively, in turn. Hence, it is vital that you think imaginatively as well as interpretively when you encounter difficult problems. By the use of brainstorming you may find that what you thought is a problem really isn't one after all. Or, you may find that it is only part of the problem. Hence, the hypothesis should be changed or modified. You may find that other ideas should be considered, as a result. Through brainstorming you may discover neglected sources to be tapped, areas to be investigated, subquestions to be answered, factors to be analyzed. You may find ideas for experimentation and observation—ideas for apparatus, materials, and procedures; additional methods of research; and standards for evaluation along with other determinants.

The evaluation of ideas depends on the individual problem being studied. However, several general standards apply to all research problems:

1. Is there *precedence* for the idea? Has it, or something like it, ever been done before?
2. In light of the nature of this problem, is the idea *feasible*? That is, can it be done?
3. Is the idea *practical*? That is, is it worth doing?
4. What is the probable *result* of doing it? How will the people who are involved be affected?

At times, you will find, one more question is needed as a general standard: does the idea have *appeal*? That is, is the idea aesthetically pleasing as well as worthwhile?

Naturally, in addition to using these, you will always want to isolate the particular standards that apply to your specific problem.

One final note of caution is in order. Your imagination *can* play tricks on you. For example, you might in brainstorming produce an idea that is contradictory to a known truth. When you evaluate the idea, you must be careful not to replace what has been proved with an idea that has not.

EXERCISES

FACTS AND THE MEANINGS OF INTERPRETATION

1. Classify the statements below as one or more of the following: a fact, a conclusion, an evaluation, an idea.
   a. A square building lot is needed for this particular house plan.
   b. The architect said, "You need a square building lot for this house plan."
   c. A square is a four-sided figure whose lines are equal in length and at right angles with the adjacent lines.
   d. Perhaps a square building lot is best for this particular house plan.
   e. A square lot is the best shape for a building lot.

2. Which of the statements in Exercise 1 can be stated (with changes) as hypotheses? Discuss you answer in each case.

RESEARCH

3. Prepare a brief report on the pure research you have done as a student (or as a part- or full-time employee).
4. Prepare a brief report on the applied research you have done as a part- or full-time employee (or as a student).
5. Using all sources of information available to you, prepare a list of the indexes, abstracts, dictionaries, etc., in your specialized field of study. If necessary, call on librarians, professors, etc., for help.
6. For a particular topic you have chosen to research and report on, prepare a list of *all* sources (including books, articles, and pamphlets) of information you may be able to use. If necessary, call on librarians, professors, etc., for help. (To make sure that your material is up-to-date, take at least half of your sources from publications issued within the past year.)
7. For a research project that you are beginning or are about to begin, think about and be prepared to suggest how you may be able to make use of
   a. observation (in the broad sense).
   b. personal interview(s) with one or more authorities on your subject, such as local professors, engineers, businessmen, and government officials.
   c. one or more letters of inquiry to an authority on your subject who is too far away for a personal interview.
   d. a mail questionnaire to members of the population who are concerned about, perhaps even somehow involved with, your subject.
   e. pretesting.
   f. sampling (representative or weighted).

NOTE TAKING AND DOCUMENTATION

8. Find and quote excerpts of published writing that include a quotation, a paraphrase, and a summary.
9. The following footnotes from a report are complete and in the correct numerical order. However, the elements of each footnote have been arranged haphazardly. Arrange the elements so that all footnotes are presented as shown on pages 80–81.

```
1  "Precision Alignment Systems" by Kale Skutely. In Civil
   Engineering for September 1967, page 286.
2  From Proceedings of the 31st American Congress on Survey-
   ing and Mapping. "Laser Ranger--the New Distance Measur-
   ing Instrument for Surveyors" by Harry R. Feldman. Page
   504. March 1971.
3  In Civil Engineering for February 1971: "Short Range EDM:
   Big Boom for Surveying." Page 142. By Eugene E. Dallaire.
```

4 <u>Lasers</u> <u>and</u> <u>Their</u> <u>Application</u>, page 1. By Kurt Stehling.
   Cleveland, 1966. World Publishing Company.
5 The same as footnote 4.
6 The same as footnote 3.
7 The same as footnote 4 but taken from page 2.

10. Prepare a bibliography of the four sources named in Exercise 9.
11. Using the brief-documentation and references system, present the footnotes listed in Exercise 9 as they would appear in the text itself.

LOGIC AND ILLOGIC

12. In another of your textbooks, in a laboratory report, etc., find and record two or more closely related facts. On the basis of these facts, draw one or more conclusions inductively. Then draw one or more conclusions deductively. Identify clearly each type of thinking you have used.
13. Look carefully at the following "problem statements." Are there any weaknesses? Where are they? Be prepared to discuss and to improve the statements. Make any assumptions about the nature of the individual problem that you think are necessary, but be sure to report these assumptions.
    a. Air pollution at the Hagen Coal Company.
    b. What can be done about heterosis at Big Spring Farm?
    c. Acid mine drainage.
    d. Processes for recovering crude oil.
    e. Is water desalination feasible?
14. What "illogic trap" is apparent in each of the following excerpts from reports?
    a. "A large segment (52%) of the people interviewed favored this change in the design of our pyrometer."
    b. "In the first test, 1.8 gallons of gasoline were consumed each 50-mile run; in the second, 1.6; and in the third, 1.7. These results prove that the new carburetor will ensure owners at least 31 miles to the gallon."
    c. "Dr. James E. Prentice, Director of Research at Miles Astronuclear Laboratory, stated that the town-manager form of government would be more efficient than the form we have now."
    d. "The results of more than two years of continual testir₃ indicate a possibility of 100% efficiency. Not once was a lower value recorded."

SPECIAL USES OF INTERPRETATION DURING INVESTIGATIVE WORK

15. Select a subject on which you could write (a) a problem-solving report, (b) an evaluation report, and (c) a justification report. Then be prepared to discuss in class how you would develop, in detail, each of these reports.
16. Discuss the effectiveness or ineffectiveness of the following evaluation by standards.

Evaluation of Metals from Lance Construction

In the basic oxygen process, a mixture of oxygen and natural gas is blown into the bath of liquid metal. The mixture is introduced to the molten metal through a long water-cooled pipe called a lance. The purpose of this report is to evaluate the metals that can be used in the construction of the lance.

The standards for the metal to be used in the lance are its ability to withstand high temperatures, its small coefficient of linear expansion, its ability to be welded, its high thermal conductivity, its chemical inertness to all components of the bath, and its inexpensiveness.

Three metals that are acceptable for lance construction will be compared: stainless steel (Type 304), carbon steel, and copper.

If the highest melting point is considered the best, stainless steel would be the best metal.

As far as the coefficient of expansion is concerned, all three metals are within tolerable limits.

The three metals can all be welded with varying degrees of success.

Concerning the thermal conductivity, copper is the best of the three metals.

Under operating conditions, all three metals would be relatively unreactive to all the compounds that contact the lance.

Cost, which must be relatively low, is often the deciding factor in the selection of a metal. Not only the initial cost but also the cost per day must be compared for all three metals. For example, a lance that would last indefinitely could be made out of platinum. However, the cost would be much too great. It is cheaper to make 100 copper lances, which won't last as long, than to make one lance out of platinum.

Copper is therefore recommended for lance construction over the other two metals examined.

IMPORTANCE OF STANDARDS IN ALL INTERPRETATIONS

17. Examine a recent experience you have had in the laboratory, in the library, in the classroom, at work, in a store, at home, etc. What standards come to your mind as you think about the details of that experience? In sum, how well were the standards met? What standards in particular were not met fully? What standards in particular were not met adequately? Discuss fully your answers to these questions.

IMPORTANCE OF USING IMAGINATION IN INTERPRETIVE WORK

18. Alone or as a member of a group, practice brainstorming by suggesting new uses for one or more of the following: a paper clip, a man's old felt hat, an old paint brush, a discarded toothbrush, an ordinary red clay brick, an old wet (or dry) mop. Assume that you have as many of each item as you wish and that you can do anything you want to the item (bend, cut, twist, crumble it, etc.) in each case.

SPECIAL PROBLEMS

19. For research you are now doing, present a report listing the titles of and summarizing the content of at least five secondary sources of information (books, journal or magazine articles, newspaper articles, etc.) on your subject. Write at least 50 words for each summary, and indicate the use you now plan to make of each source.

20. For research you are now doing, present a report of 150 words or more on observations that you have made over a specific period of time (in the laboratory, in the field, on the job, etc.).

21. Draw up a list of interview questions that you would like to ask of a local authority on the subject you are now researching.

22. Write a letter of inquiry to one or more authorities on your subject who are too far away for a personal interview. (See pages 157–61 for suggestions on writing an effective letter of inquiry.)

23. For research that you are now doing, prepare a questionnaire for a population somehow affected by the subject of your study. To save postage, reproduction, and other costs, take the questionnaire to each member of the population (or to a sample of it) and administer the questionnaire face to face.

24. For research you are now doing, develop an analogy to help your reader understand more easily your subject or some aspect of it.

25. Prepare a report on your experience in using brainstorming during a current research project. The report should be presented in three parts: (1) a list of the ideas that you thought of *before* you tried brainstorming, (2) a list of the ideas suggested by a group you gathered to brainstorm your subject, and (3) your evaluation of each of the ideas suggested by your group.

26. When you complete your research on the subject you are investigating, prepare a rough outline of the body of your report written to complete the project. Submit the outline to your instructor in the form he requests.

27. Bring to class, for your instructor's review, your file of note cards containing quotations, paraphrases, and summaries of sources that you have used for information. Arrange your cards in the order you now plan to use them in developing your report on the investigation.

Chapter Six

DEFINITION
AND DESCRIPTION

To help leaders make decisions, your reports must contain essential information —about efficiency, cost, service life, maintenance, etc. Often, however, even more basic information should be shared.

Suppose, for example, that you are assigned the problem of finding a way to improve the efficiency of your company's incineration of waste materials. But in your research you learn that little can be done with your present equipment. Thus, you recommend the purchase of a new, different incinerator. This incinerator will burn up not only waste material but also grit, noxious gases, and even the smoke itself.

Naturally, the decision maker who reads your report is interested. "An incinerator will do all that?" he asks. He is not skeptical, of course. He has faith in your competency as a researcher and an evaluator. And the essential data are right in front of him: your report thoroughly covers efficiency, cost, etc. He has the information on which to base a decision. But he is curious and he can't help asking other questions: "What exactly is it? What does it look like, and how does it differ from old-style incinerators? What are its component parts and how do they function?"

This decision maker is asking the questions that are answered by the most basic kinds of information—definition and description. *Definition is a statement of the meaning of something. Description is the presentation of a mental image in words.* Definition tells the reader what, basically, something is in itself—or what it is in contrast to something similar. A lathe is one kind of power shop tool. Perhaps to define it, a writer will briefly explain its purpose; or, if he thinks that is not enough, he will show also how a lathe differs in function from other power shop tools. Description goes a step further. Since description is a word picture, it helps the reader to visualize the lathe.

In many situations, the reader needs both definition and description. For example, the reader in the case cited earlier knows nothing about the new incinerator. Thus, he wants to know what, precisely, a smoke-burning incinerator is. He also wants to know what it looks like, what its parts are, and how it works.

This example may seem dramatic, but it is not unique. Since no two people are exactly alike, the experiences and backgrounds of the reader and the writer never overlap completely. Indeed, two people may be so unlike that a writer's terms are unknown to the reader. Obviously, communication cannot take place in such a situation. Hence, a writer should never take for granted what a reader does and does not know. Instead, he should use the following as a rule of thumb: if from previous experience he *knows* that his reader speaks the same language, he can ignore definition in his writing. If he does *not* know, he should include it. The same can be said about description.

DEFINITION

Definitions vary in length from a word to a treatise, depending on how complex the subject is. But in all definitions the influence of a characteristic structure is apparent. This structure is called a *sentence definition.* A sentence definition consists of a *term*, the thing to be defined; a *genus*, the class of things to which the term belongs, and a *differentia*, the feature of the term that sets it apart from other terms in the same class. Following are simple examples. Note that all the terms are in the same class—instruments used in meteorology:

A thermometer (term) is an instrument (genus) that measures temperature (differentia).

A barometer (term) is an instrument (genus) that measures atmospheric pressure (differentia).

A hygrometer (term) is an instrument (genus) that measures humidity (differentia).

Note also the basic structure: A _____ is a _____ that _____. Some writers prefer to use "The" instead of "A" and "which" instead of "that." Then they can easily remember the basic parts of the structure (The _____ is a _____ which _____) by having memorized them as "Thee is a witch."

Of course, the basic structure is varied. For example, if you turn to "lathe" in the Funk & Wagnalls *Standard College Dictionary*, you will find this entry:

lathe (lāth) n. 1. A machine that holds and spins pieces of wood, metal, plastic, etc., so that they are cut and shaped when the operator holds abrading tools against them.[1]

[1] Funk & Wagnalls *Standard College Dictionary*, New York: Harcourt Brace Jovanovich, Inc., 1963, p. 764.

Even here the basic structure is indicated. All you have to do is add the article *A* or *The* at the beginning, and then the verb *is* after *lathe.*

Sometimes in going to the dictionary, you will find only a phrase or a single-word synonym (equivalent). For example, under "barometer" the basic definition is "an instrument for measuring atmospheric pressure."[2] In your own writing you can quote the phrase or present a complete-sentence definition, as you prefer. You can also combine definition with the development of your thought. If you are writing instructions, for instance, you can say, "Check the barometer for the atmospheric pressure." If you are reporting results, you can write, "The barometer showed the atmospheric pressure to be very low."

Frequently, a one-word synonym is enough to make a term meaningful to your reader. In Chapter One of this book, "analogy," a key word, is very simply defined as "similarity." Since a long analogy was being developed there, neither a complete sentence nor a phrase was necessary. By carefully structuring your development of a report, you can make many of your terms self-explanatory.

Most writers shy away from defining because they fear that they will insult the reader's intelligence. Have no such fear yourself. If you think your reader may be insulted by a formal structure, try using a phrase or synonym, as recommended above. In using either, you can make the meaning clear by your development—and you will insult no one, because the definition will be informal.

Qualification of the Term, Genus, or Differentia

The illustrations of simple sentence definition have one quality in common. In each, the term, genus, and differentia are all very brief. Generally, your definitions can be as simple as these. However, qualification is needed when the term is used with a limited or special meaning, when the genus is too broad, or when the differentia is inadequate.

Qualification of the Term

For example, the writer may have to restrict the meaning of a key term. Note that in each of the following illustrations, a key term is used in a very special way:

```
The word "distillation" here actually means partial distilla-
tion; that is, heat is applied only until separation begins.

As used in this study, "efficiency" means the amount of input
that can be recovered later as a useful product.
```

You can see that qualification of the term may be important. Unless he is told

---

[2]*Ibid.*, p. 116.

otherwise, the reader understands a term to mean what the dictionary says it means. More than one report writer has been reprimanded for using a term in a special way without qualification.

Qualification of the Genus

Often a genus can be simply and broadly stated—as it is for the meteorological instruments defined on page 99. No problem arises as long as each of the three parts is simply stated. But in some definitions the genus is stated very broadly, and then the burden of explanation falls on the differentia.

Take, for example, this student definition of a tellurometer: "A tellurometer is a device that can be carried around to measure precisely the phase length of various electromagnetic waves." Here, both the term and the genus are covered in five words, whereas the differentia requires fifteen. How much more effective is this revision, in which the qualified genus carries its share of the load: "A tellurometer is a portable, precise measuring device that determines the phase length of various electromagnetic waves." Now the term and the genus are covered in eight words, and the differentia is reduced to nine.

More important than the balance is the point that the genus is so restricted that differentiation is made easier. There are thousands of devices in the world, but some are eliminated at once by the use of "portable"—and many, many more by the use of "precise" and "measuring."

Qualification of the Differentia

Even when the genus is properly restricted, the differentia may be quite long because it is necessary to include several distinguishing features. In such a case, we realize immediately that we are no longer talking about *a* differentia; we must talk about differentia*e* (the plural of differentia).

Complex machinery and electrical systems are sometimes so involved that simple differentiation is impossible. Often it is best to define such devices and circuits by a short, generalized definition followed by a sentence that includes specific differentiae. Here is one example, a definition of a multipurpose control panel:

```
    The 18LN5 control panel is the electrical brain of the SCM
system's components which automatically initiates and regulates
major car operations. Upon receiving signals from voltage- and
current-measuring devices in the control and traction-motor cir-
cuits,the panel performs such jobs as (1) setting the car rate of
accelerating and braking, (2) regulating control-circuit voltage
and current, and (3) limiting load current to a safe value.
```

The clause "which automatically initiates and regulates major car operations" is a little too generalized to be truly meaningful. Hence, for more precise differentiation, the writer has added a sentence with particulars on "initiating" and "regulating."

Obviously, the generalized differentia should be composed with great care. As in the example, it must always prepare the reader for the specific differential in the sentence that follows.

Use of Figurative Language and Analogy

In the first sentence of the example just cited, you may have noted the word "brain." "Brain" here is figurative (i.e., not literal) language. Specifically, it is a *metaphor*—an implied, but not stated, comparison. The panel is not a real brain, of course. However, it functions as one in initiating and regulating the many automatic controls. Figurative language is always helpful if a qualification, like the second sentence in the illustration, is added to make it meaningful.

Analogy, the comparing of things that are essentially unlike, also helps one to define. Consider the following example of its use. Students of electrical engineering and related subjects have no difficulty with a word like "saturated." However, others may be unsure of its exact meaning. Analogy will help the latter to understand completely; although electric current and water are unlike in all other respects, both are said to "flow":

```
If no more magnetism can be forced through a piece of iron, the
iron is said to be "saturated," or filled. To further increase
the amount of magnetism is like trying to force more water through
a hose than the hose can carry.
```

Pitfalls in the Writing of Definitions

When you write sentence definitions, be sure to follow the form faithfully. If you do not, the result will be confusion, rather than clarification, for your reader. Listed below are common weaknesses and errors that create confusion.

1. Appearance of Definition

Form only—incomplete: A motor-generator set is a combination motor and generator which are built into one machine.
(This statement is "circular"; that is, the part of the sentence that follows the verb "is," is merely a restatement of the term!)

Complete definition: A motor-generator set is a mechanically connected, voltage-producing unit which satisfies the current requirements of the battery and the control equipment of a rapid-transit car.
(This statement is a complete sentence definition, containing a restricted genus, "a mechanical ... unit," and a meaningful differentia, "which ... car.")

2. Wrong Grammatical Structure for the Term or Genus

Illogical: In electricity, *saturated* is the filling of . . . .
("Saturated" is a verb form used incorrectly here; it can be used as an adjective but it cannot be used as the equal of a noun—as a subject. "Filling" is a verb form used as a noun.)

Logical: In electricity, saturation is the filling of . . . .
("Filling" is now correct because the term, "saturation," is a noun, also.)

Of course, other grammatical structures can be used for the term and the genus:

Logical: In electricity, to saturate is to fill . . . .
(Both "to saturate" and "to fill" are infinitive forms of verbs.)

Logical: In electricity, saturating is filling . . . .
(Both "saturating" and "filling" are gerunds in this use—verb forms functioning as nouns.)

3. Ungrammatical Verb

Illogical: In electricity, saturation is when (is where) . . . .
("Is when" or "is where" is grammatically wrong here. In this kind of structure, "where" or "when" introduces an adverb clause, which cannot take the place of a noun.)

Logical but cumbersome and wordy: In electricity, saturated is the word used to indicate a point at which . . . .
(The structure is much too long; it holds the reader off unnecessarily. A term should be explained as quickly and simply as possible.)

4. Vague, Inexact Verb

Vague and inexact: In electricity, saturated refers to (relates to, concerns, etc.) . . . .
(Verbs such as "refers to," "relates to," and "concerns" are not only vague but also inexact. "Saturated" or "saturation" does not "refer to" or "relate to"—nor does it ever "concern"—*anything.* A term always "is" something, or it "means" something: the genus always tells what kind of a thing a term *is* or what it *means.* What is in the genus is another word for what is in the term.)

5. Use of the Term in the Genus

Illogical: A resistor is a device offering electrical resistance which is used in an electric circuit for protection or control.

(To say that a resistor offers resistance is to explain nothing. No definition should repeat the term, or a form of the term, in the genus. The problem often arises because, in turning to the dictionary, you find that the root word has already been defined. Here, for example, you would find that both "resist" and "resistance" have been defined—as they should have been—because alphabetically they precede "resistor." Yet, a definition that you write should not send the reader to the dictionary!)

Logical: A resistor is an obstacle, put into an electric circuit either to drop voltage or to reduce current, which protects or controls.
(Now the reader knows what a resistor is. He would not know if he read the definition whose genus includes a variant form of the term.)

At times it may appear that you must use the term itself, or a form of it, in the genus. Take, for example, the term "force pump." In the dictionary you will likely find "pump" for the genus—hence, the temptation is great to use "pump" yourself. But if you look up the term, you will *always* find that you can use directly, or adapt, words used in the definition of the term. In this case, for instance you can write: "A force pump is a mechanism with a piston and valves that raises a liquid and forcibly ejects it under pressure." To be sure, most people think that they know exactly what a pump or other familiar item is. But if you ask, some of their answers will be surprisingly vague and inexact. In writing sentence definitions, then, you should never include the term in the genus. After all, definitions are always written for the sake of explanation and you cannot explain a term if you make it a key word in your "definition."

Extended Definition

A sentence definition is not always as much explanation as your reader wants. If the term that is the subject of your report or article is new to him, difficult to understand, or especially interesting, he will want to know more than a sentence definition tells him. If it is a picturable term, he would like to see, in illustrations as well as word description, what it looks like. If it has parts or divisions, he would appreciate knowing what they are and how they are related to one another. If it names a problem, he would be grateful for a complete definition of the problem—causes, specific details, and harmful or undesirable results. If the term has significance, he would be interested to learn what makes it worth knowing about.

These are all good reasons for adding a few sentences or paragraphs, or even several pages, to a sentence definition of the subject of your writing. The length of the addition always depends on the amount of information available, the complexity of the subject, and the extent of your reader's wants in a certain situation.

Development of Extended Definition

Since a sentence definition is a generalization, the most common order of development for extending it is from generalization to details. This is one of the ten orders of development discussed on pages 29-30. The other nine orders are (1) details to generalization; (2) time, or sequence; (3) space, or location; (4) familiar to unfamiliar, especially by analogy; (5) comparison, or similarity; (6) simple to complex, or principles, etc., first; (7) contrast, or difference; (8) cause to effect; and (9) effect to cause. Invariably, two or more of these orders are combined.

In addition to these orders for developing an extended definition, three others are especially appropriate: etymology, elimination, and illustration or examples. *Etymology* order is the historical tracing of a word's origin and meaning(s) since the word came into use. In technical writing, definition by a review of how a word came into use often helps to explain its precise modern use. *Energy*, for example, is used in modern physics to mean the capacity for doing work and for overcoming inertia. In the Late Latin period (200-600 A.D.), *energy* meant essentially what it means today: vigor of action. In the original Greek, however, the word meant simply "at work." Of course, it is the original Greek use that is more accurately identified with the scientific use today.

Elimination order is the development suggested by the word itself. Possible meanings are eliminated, one by one, until the writer—in a final sentence definition—indicates exactly how he is using a particular word. This order is extremely helpful in distinguishing the use of a word that has many meanings. *Laminate*, for instance, has four similar meanings: "(1) to beat, roll, or press (metal) into thin sheets; (2) to separate or cut into thin sheets; (3) to make of layers united by the action of heat and pressure; and (4) to cover with thin sheets or laminae."[3] When you write a report, it may be very necessary to show that you have only one of these meanings in mind. If the meaning is that of (4), you will do well to show that laminate, as you are using it, is not (1), (2), or (3).

Illustration or examples order is the development that makes abstract, or generalized, definitions meaningful by the use of specifics the reader will readily understand. Although illustration is rarely used in technical writing, a story or an anecdote may be the best way of extending a definition. For instance, if you are writing a report on the "cracking" (distillation) process in petroleum manufacturing, you might well explain by telling how "cracking" was accidentally discovered in 1861.

More common, and more helpful in reports and articles, is the use of short, specific examples. A sentence definition like "Marine farming is the cultivation of salt-water organisms which affords man a source of edible food" may mean little. But if you add examples, such as "the suspension of young oysters from

---

[3]*Ibid.*, p. 758.

floats" and "the raising of ocean fish in salt-water ponds," the definition becomes meaningful.

An important point about the use of these thirteen orders is that several, perhaps many, will be helpful in a long definition.

The Writing of an Extended Definition

The quotation of part of an extended definition may help you to plan your use of this important technique in your own writing. The following excerpt is from the introduction to a report entitled "Recognition of Meteorite Impact Sites by Analysis of the Effects of Shock Metamorphism." This report was written by a college senior majoring in geology, but it could have been written by a surveyor for a mining company looking for ore bodies deposited in fractured and deformed rocks.

> The well-formed meteorite craters on the moon have been the backdrop for portentous events in recent years. Only within the last decade, however, have geologists declared with certainty that Earth also bears scars from her own fair share of impacts by large meteorites. Millions and hundreds of million years of erosion by water, wind, and ice sheets, and the upheavals associated with mountain building, may thoroughly obliterate the distinctive circular structure of a meteorite crater. However, the site of impact can still hold ample evidence of the catastrophe that once occurred there. Five types of such evidence will be described in this report.
>
> A meteorite of several thousand tons traveling at twenty miles per second at the time of impact will not only punch out a crater but will also cause changes in nearby rocks and minerals. Crystals of the common minerals quartz and feldspar often develop planes of weakness. Sometimes whole crystals are found with their formerly rigid internal arrangements of atoms shocked into complete chaos. On a larger scale, sections of the original rocks may melt at the time of impact and later solidify into new rock. Unusual funnel-shaped fracture surfaces, called shatter cones, often form. Such results of the instantaneous transmission of a high-pressure shock wave through the rock, analogous to passage of an earthquake or seismic wave, are called shock metamorphism.

The introduction continues with a paragraph on the value of doing research on shock metamorphism. Then the report describes the five types of evidence—as the writer indicated he is going to do, at the end of paragraph 1.

An analysis of the two paragraphs that are quoted helps us to understand the nature of extended definition. The first paragraph thoroughly introduces the subject, shock metamorphism, and the second paragraph thoroughly defines it. All the writer needs to do in the rest of the report is to elaborate on the differentiation—the five types of evidence.

Note that the definition is not completed until the end of paragraph 2. The first 24 words of sentence 1 of paragraph 2 anticipate the genus ("the instantaneous transmission of a high-pressure shock wave through the rock") in the last sentence. The rest of sentence 1 anticipates the differentiae (note the plural), which are presented in sentences 2, 3, 4, and 5. These sentences cover all five types of evidence. The term itself is the last two words of the paragraph.

Of course, the writer *could* have begun with the term, "shock metamorphism." Then, however, he would have to reorganize the two paragraphs completely. In this particular case, obviously, the development should build to the definition rather than depart from it. Most of the time—as you might expect—it is better to start with the definition. A skillful writer, like our author here, can use such a development when it is appropriate, as it is here. But a less-experienced writer will do well to put his definition at, or very close to, the beginning of hs report.

Analysis also helps us to see how several orders of development can be used together. Sentences 1 and 2 of the first paragraph quoted establish a sharp contrast: craters on Earth are contrasted with those on the moon. Sentences 3 and 4 establish another contrast: there may be no evidence of craters visibly, but the site may hold "ample evidence" nevertheless. Sentence 3 is also an example of cause-to-effect development. And the last sentence illustrates details ("five types of evidence") following a generalization ("ample evidence") and a sentence that is even more generalized—sentence 2 (with its "also bears scars"). Sentence 2, it should be apparent, is the topic sentence of the whole paragraph.

Paragraph 2 is an excellent example of two orders of development—cause to effect and details to generalization—that are sustained. There is even a third order, although it ends before the last sentence. The first sentence is a generalization supported by sentences 2, 3, 4, and 5, which point out specific details of "changes in nearby rocks and minerals."

These notes should help you to write effective extended definitions. You should realize, however, that no defining task is ever either simple or mechanical for the skillful writer. Developing a really good definition report or article is always a challenge—as the case study just presented clearly demonstrates.

CLASSIFICATION AND PARTITION

Before we leave the excerpt on page 106, we should note that the illustration of extended definition also illustrates the use of two adjuncts to effective technical writing. These adjuncts are called *classification* and *partition.* Classification is the listing or arranging of *things of a kind.* Partition is the listing or arranging of the components or characteristics of *an individual thing.* In the excerpt, classification is apparent in the listing of *five* types of evidence; partition, in the indication that *each* of the five will be described. A classification is always made of two or more things; a partition, of an individual thing.

A classification or a partition may be broad or narrow. A broad classification of books, for example, is into trade books and textbooks; a narrow classification is into textbooks for a course in technical writing. A broad partition of a book is into its physical parts: cover, binding, and pages; a narrow partition is into its contents: all the topics covered in, say, *this* book.

Whether broad or narrow, every classification and every partition indicates a single point of view. A classification is made on the basis of something. A partition is made according to how the partitioner views the whole individual subject he has selected. The broad classification into trade books and textbooks, for instance, is made on the basis of why they were written primarily—to entertain the general public or to instruct students. The narrow partition of a textbook on technical writing depends on how the author views the teaching of the subject. The point of view always indicates what either the classification or the partition covers in each case.

The point of view should be clearly indicated before a classification or a partition is made. In classification, the phrase "on the basis of" or a similar phrase (such as "according to") should appear in order to show the logic of the division. Frequently, an outline helps the classifier to be sure that his division is logical; the broad classification of books, for example, is presented in outline as follows:

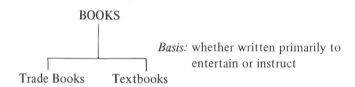

BOOKS

Basis: whether written primarily to entertain or instruct

Trade Books   Textbooks

If a classification is illogical, *cross-classification* or *overlapping* should be apparent. Cross-classification can be seen in an attempt to divide books into trade books and first editions; trade books and first editions are not of a kind, and thus no basis can be stated. Overlapping can be seen in an attempt to divide books into textbooks and textbooks on technical writing; textbooks and textbooks on technical writing are not equal, for the former include the latter.

Overlapping classifications indicate that more than one division—that is, a classification and one or more *subclassifications*—should be made, each successive logical division being a narrower one. Trade books, for example, can be subclassified into fiction and nonfiction on the basis of whether the work is primarily either imaginative or factual. And so on.

Logic and illogic in partitioning are discussed and illustrated on pages 32–34; an outline is the ideal example of a partition. While there is no such thing as a subpartition, the division of a single subject into component parts *can* be either quite broad or extremely thorough and detailed. The partitioning of this book, for instance, can be broad (into chapter titles only) or thorough (chapter titles, first-degree headings, second-degree headings, etc.).

Of course, in the final analysis, both classification and partition are only lists and arrangements—"skeletons," we might say. What is needed is "flesh" to give them substance: all the qualities and techniques that are covered in Parts I and II.

DESCRIPTION

Description is always definitive. By words (and pictures when they are needed or desirable) the writer creates a definite image of his subject in the reader's mind. The definition enlarges as the description develops, even though the writer's announced intention may be only to "describe."

In technical writing, description takes two forms: "device description" and "process description." Both are very valuable techniques for any aspiring technical writer to master.

Device Description

In technical writing, we tend to think of complex things—machines, elaborate apparatus for experiments, control panels, etc.—when we think of the word device. It is quite natural to do so. Most of the devices in science and industry that need to be explained *are* complex. In fact, however, we can say that anything that serves a useful purpose may be called a device. This book, the chair you are sitting on, the lamp you are reading by—these and thousands of other items are, therefore, devices.

It is important that we define "device" broadly. Some of the simplest, least-complex-looking things often need to be explained. The thermocouple, bellows, dew gauge, disk harrow, blowtorch, and reverberatory furnace are all examples of simple devices. But if one of these, or a similar device, is the subject of our writing, use of the word without definition and description may disturb our reader. If different designs of a dew gauge are being compared, for example, descriptions—with photographs, drawings, or at least sketches—of all designs should be included. Then the reader can clearly see the differences that are being pointed out.

Outline for Describing a Device

A logical plan is always needed for describing a device. An outline that the reader's mind can follow easily is as follows:

Definition
Statement of significance
Underlying principle
Generalized view

List of main parts

Analysis of each main part

Summary of the device's operation or function

Let's examine this outline more closely as a set of guidelines for you to follow.

Definition Always include a complete sentence definition, preferably at the very beginning, unless you *know* that the reader is familiar with the device:

```
A relay is an electromagnetic, remote-controlled switch which opens
and closes one or more electric circuits.
```

Statement of significance Add a statement of significance when the definition does not fully indicate the importance of the device:

```
The relay opens circuits so that one or more other electric devices
will operate in sequence, and closes circuits to stop operation or
to protect against electrical failures.
```

Underlying Principle Whenever necessary, note the underlying principle—the natural or scientific law—on which the function or operation of the device is based:

```
To function, the relay makes use of variations in the strength of
the current it receives.
```

Generalized view Always give the reader a generalized view, so that he can picture the device clearly. Vital to the reader's understanding is a clear statement of the position from which the device is being described (top, side, rear, etc.). It goes without saying, that a photograph or drawing showing this point of view and including the names of major parts, will be very helpful.

Sometimes, a photograph or drawing can take the place of a generalized description in words. However, even a brief orientation in your text—perhaps by analogy—will give the reader a view that he can identify with his own experience. (Some photographs and drawings give no meaningful indication of the size of what is shown.)

```
In appearance, the tellurometer closely resembles a large, portable
radio. Indeed, a glance at the face may make one think that he is
looking at the front of a table model.
```

List of main parts Include a list of the main parts of the device, so that the reader can know in advance what is going to be described in detail:

```
The mulcher, which is mounted on a trailer and powered by a 56-
horsepower V-4 Ford engine, has five major parts. These are the
tank, pump, nozzle, and two sets of grinding blades.
```

Analysis of each main part   Analyze—that is, break down into components—the parts of each main part. In so doing, follow some logical order in space: top to bottom, front to back, clockwise, etc. Be sure to tell the reader which order you are following. Also tell him what each part is, what it looks like, and what function it performs. If your transitional words and phrases do not indicate the relationship each part has to one or more other parts, supply a sentence so that the relationship(s) can be readily understood:

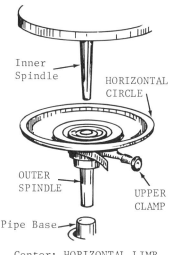

The middle part of the transit is the horizontal limb. At the center of the horizontal limb is a vertical pipe, flared at the top and tapered at the bottom, which is called the outer spindle. Its flared end holds the inner spindle [of the transit's upper part], and its tapered end fits into the pipe base in the lower part.

Flush with the top of the flared end is the center of a saucer-like plate. This plate, called the horizontal circle, has a graduated outer ring on which horizontal angles are measured. Just below the plate is a screw-clamp, called the upper clamp, which when tightened will prevent the inner spindle from turning.

Center: HORIZONTAL LIMB
    OF THE TRANSIT

Figure 1

TRANSIT SUBASSEMBLIES

Summary of the device's operation or function   Last, give the reader a brief but comprehensive summary of the device as an operating or functioning unit. Since your reader is by now familiar with the parts, at least the names of the main parts should be used freely. The purpose of this summary is to help the reader understand how what he has been reading about works:

When the lead weight at the one end of the string rises or falls, the tension on the string decreases or increases. The float on the other end of the string then rises or falls accordingly. The amount of tension on the string causes the pulley to turn, and the stylus at the arm's tip fluctuates upward or downward on the chart. Thus the level of the liquid in the standpipe, and hence the weight of the snowpack are recorded on the revolving chart. Since the chart rotates at a constant speed, the time of each fluctuation is known exactly.

Role of Definition Throughout

Notice that definition is used wherever necessary or helpful in the description of a device. Both the statement of significance and the statement of underlying principle add to the sentence definition of the subject. Presenting a generalized view and listing the main parts help the reader to picture what he is reading about; they also help him to understand the photograph or drawing quickly. Definition of parts, both major and minor, is included whenever their names are not self-explanatory.

Process Description

A process is any series of actions in a sequence that bring about a particular result or condition. Studying is a process. So is driving a car, getting dressed or undressed, or doing a laboratory experiment. Thus, we see that a process must be defined as broadly as a device. And fairly simple procedures, techniques, methods, activities—as well as complex ones—must be included when we speak of processes.

A process can be described in one of two ways. We can describe a series of continuous actions from the viewpoint of an *observer*, or onlooker, on the one hand, or from that of a *participant*, or doer of actions, on the other. We would write from the viewpoint of an observer if we merely wanted to inform our reader. We would write from the viewpoint of a participant if we wanted to help our reader perform the actions himself if he wished. Clearly, we should confine ourselves to only *one* of those viewpoints; to mix the two would be to confuse the reader completely.

Regardless of which viewpoint you write from when you describe a process, include illustrations. You should have not only photographs and drawings, but also flow charts, diagrams, or schematics as needed to show individual actions in sequence. Naturally, if you are writing for observers, drawings and photographs, etc., do not have to be extreme close-ups. However, if you are writing for participants, you will need close-ups whenever readers need precise details on what to do and where to do it. Observers do not need to know exactly, but participants *do.*

Since you must never mix the two viewpoints in describing a process, we shall examine them separately.

Outline for Describing a Process for an Observer

The reader who is simply an observer will find the following plan of development meaningful:
Definition
Statement of significance
Underlying principle

List of major steps
Analysis of each major step
Appropriate ending (if needed)

Let's look at this outline more closely.

Definition   Unless you know that your reader is familiar with the process, always include a complete sentence definition, preferably at the very beginning:

Steam distillation is a chemical-physical process which separates two substances in liquid solution that have different boiling points.

Note that the introductory article (a, an, the) is not used for a process term that does not include it ordinarily.

Statement of significance   Work in a statement of significance when the process being described has special importance. Sometimes, as in the following example, it may be necessary to present this statement before the definition:

The great advances in technology in the past few years have produced many results, not all of which are desirable. Among the results is the presence of difficult-to-remove compounds in our rivers and streams. To rid our waters of these harmful materials, we need to develop new ways of treating waste water. One such method that we already have is called flocculation.

Underlying principle   Also try to insert a statement of the underlying principle—the natural or scientific law—on which the process is based. The introduction to the report cited above, for example, notes that Stokes' Law is the basis:

The basis of the flocculation process is Stokes' Law. This law states that the larger a particle becomes, the faster will be its terminal velocity, or the rate at which it falls.

List of major steps   Include a list of the major steps, or phases or stages, of the process, so that the reader will know in advance what is going to be described in detail:

The operation of the cathode-ray oscilloscope covers four distinct phases: emission, horizontal deflection, vertical deflection, and screen tracing.

Analysis of each major step   Report all of the activities that take place during each major step of the process:

The final stage in pulpwood harvesting is readying the wood for shipment. When the cradle is full, the binder chain is tightened around the logs, and the load is dropped from the cradle to the ground with the logs still on end. The vertical bundle thus formed is bound with wire or a reusable strap, and the binder chain is released. Then, the bundle is lifted by the loader and placed on a trailer frame. When the frame is full, it is loaded on a trailer for delivery by truck to a wood yard or a railroad siding.

Appropriate ending (if needed) Although the description of a device ends with a departure from the basic job of describing parts, the description of a process may well end with the completion of the last step. Thus, a special ending is not a "must" in this kind of writing. At times, however, it may not be enough to stop as soon as the last stage of the process has been covered. For example, you may have something of significance to add to your statement—something that would have broken the smooth development of your introduction. You therefore save the information for a short, final paragraph—the only place where it will logically and meaningfully "fit in":

> According to the manufacturer of the boom apparatus used, this process is worth experimenting with because maintenance costs are very low. Even more important, the manufacturer claims, the process can be carried out in adverse weather and in a variety of sea conditions.

Role of Definition Throughout

As in the description of a device, you should define where necessary throughout the description of a process for an observer. You should note, however, that the challenge to do so, unobtrusively, is greater in a process analysis. When you describe a device, it is a simple matter to stop with the naming of a part and insert a complete sentence definition of it. When you describe a process, though, such a sentence can be very obvious because you must interrupt a smooth-flowing presentation of steps in a sequence.

The way to make such interruptions unobtrusive is to work the definitions into your description so that they appear natural. In the following example, you will notice, two terms—"tree feeder" and "delimber"—are defined in the same sentence:

A tree feeder on the tractor carriage pulls the tree, butt first, through a delimber, which strips off the branches as the tree is pulled forward.

Of course, these definitions are not complete. You don't know the genus of either term; that is, what kind of a machine each is. You also don't know what a tree feeder or a delimber looks like. However, you *do* know what each part does as an agent in the process. And you are probably not distracted, as you might be if the writer had stopped after each term and inserted a complete sentence definition ("The tree feeder is . . . ." and "The delimber is . . . .").

To be sure, more information should be given about the two terms in this example. Observers *are* curious about what kind of a thing a tree feeder is, and what it looks like. Perhaps in this case either a photograph or a drawing of each will be enough to satisfy their curiosity. If it will not, either a device analysis earlier in the description, or a separate list of definitions of all parts that do not explain themselves, will do the job.

Outline for Describing a Process for a Participant

As similar as the outlines are for writing for the two viewpoints, you should note that there is one major difference. When you write for a participant you must be precise. A participant is going to repeat the experiment you are describing. Or, he is going to overhaul an engine, exact step by exact step, as you instruct him to do. Precise details on what equipment and materials to assemble beforehand, how to proceed, what to look for, what tolerances can be allowed, etc., *must* be given.

A difference in orientation, required because of the need for preciseness, will also be apparent. When you write for an observer, you always use the third person: *he, she,* or *it* does something—or something was done by *him, her,* or *it.* In technical writing, as you know, the "it" orientation is most common:

```
A tree feeder [it] pulls the tree . . . .
```

or

```
A tree is pulled by a tree feeder [it] . . . .
```

The "it" orientation is also used in writing for a participant, but notice the difference between these two excerpts from process descriptions. The first came from a report for an observer; the second, from one for a participant:

Orientation for an observer:

```
Ionized gas molecules passing between the wire and the tube collide
with the flue-dust particles, and the particles are thereby given
either a positive or a negative charge.
```

Orientation for a participant:

```
By means of a portable vacuum pump the gas sample is pulled through
a small plug of pyrex glass wool and then through a 250-ml wide-
mouthed Erlenmeyer flask containing exactly 100 ml of standard
N NaOH and 5 ml of glycerine.
```

The orientation for the observer gives him only information. The orientation for the participant, however, gives him details so that he can do the experiment himself.

In writing for a participant, we should also note, you may have occasion to use an "I" or a "you" orientation. For example, at times you may want to show what you did *personally* to carry out an experiment:

```
I then attempted to measure, quantitatively, the amount of CO ab-
sorbed, following Schubert's method in an apparatus like that used
by Cremer.
```

Or, if you're a writer of instruction books, you will regularly have to use the imperative mood (to give a command or make a request). The reader will then see that you are addressing him directly and giving detailed instructions on how he is to proceed:

```
Measure the final pressure as is indicated in Figure 2. Insert a
thin strip of paper, wider than the tips, between the tips. Either
energize the coil to close the armature, or block the armature in
the closed position and attach the stirrup and the spring balance.
```

Other differences between writing for an observer and writing for a participant vary with the nature of the description. Explanation is included in writing for a participant if he needs to understand why a certain step or individual action is to be taken. Notice the explanations in the following brief excerpt from a laboratory report:

```
The precipitate and the filter paper are placed in a beaker con-
taining 50 ml of water and are then heated to boiling. The suspen-
sion is titrated with 0.05 N NaOH, with phenolphthalein used as the
indicator. The benzedrine sulfate precipitate dissolves rather
slowly, so that care must be taken to see that it is all dissolved.
The solution is boiled for 3 minutes at the end point to make sure
that it is not a false one.
```

Another kind of explanation is also standard in writing for a participant: a listing of equipment and materials used in experimentation. If the reader plans to duplicate the work of the writer, he will appreciate such a detailed listing. The writer may want to add actual explanations of why he used a certain piece of apparatus, say, rather than another, if the other is more commonly used in a similar situation.

Finally, but certainly not the least important point about writing for a participant, is the inclusion of *caution* and *warning* notes. A *caution note* is used when damage to a part may result from improper procedure. A *warning note* is used when an improper procedure may result in personal injury or even death. The first example following illustrates a caution note; the second, a warning note:

****
CAUTION
****

```
DO NOT STOP THE DIESEL ENGINE IMMEDIATELY AFTER A HARD RUN. IF THE
AIR AND WATER CIRCULATION IS STOPPED, HEAT STORED IN THE ENGINE
WILL BOIL THE COOLING WATER IN THE JACKETS.  IDLE THE ENGINE FOR
APPROXIMATELY FIVE MINUTES BEFORE SHUTTING IT DOWN.
```

```
*************
   WARNING
*************
```

ELECTRIC SHOCK CAN CAUSE SERIOUS OR FATAL INJURY. TO AVOID THIS
DANGER, OBSERVE AND TAKE PROPER PRECAUTIONS DURING HIGH-POTENTIAL
TESTING.

Note the use of asterisks to set off each heading from the text (and from other headings) of the process description. Note also the use of capital letters to emphasize the importance of the message in each case.

Equally important in the use of caution and warning notes is the position of the reference to the danger. Obviously, it will not do to write nothing until the warning or caution note is needed. *Reference should always be made at the beginning of the section containing the note.* (If the writer waits to present it at the moment of immediate danger, it may be too late to do anything except order a new part to replace the one that was just damaged by overheating. Or, in the more serious case, it may be too late to do anything except call the coroner.) The reference to a possible danger can be handled easily by the use of an instruction like the following; the heading indicates the beginning of a new series of instructions:

```
SHUTTING DOWN THE LOCOMOTIVE
1. Turn the page and read the CAUTION note before proceeding.
2. Move the throttle to the idle position.  . . .
```

At last we are ready to look at an outline for describing a process for a participant. Most of it is exactly the same as the one for an observer, but note the additions:

Definition
Statement of significance
Underlying principle
List of equipment and materials
List of major steps (including explanation if needed)
Analysis of each major step, with an orientation appropriate to the situation
    (including *specific* details throughout, and caution and warning notes
    if needed)
Appropriate ending (if needed)

For reminders on how to develop the definition, statement of significance, underlying principle, and an appropriate ending, review pages 113-14. The development of these parts of the description is the same whether you are writing for an observer or a participant.

EXERCISES

DEFINITION

1. Discuss the effectiveness or ineffectiveness of each of the following, which all five authors identified as sentence definitions.
   a. Kaolin is a substance used in making porcelain.
   b. A gravimeter is a type of hydrometer.
   c. Triurate refers to a fine powder or pulp.
   d. A catalyst is any substance that causes catalysis.
   e. Telodynamic concerns the transmission of power.
2. Discuss the effectiveness or ineffectiveness of the use of analogy in each of the following, which all three authors identified as definitions.
   a. Intelligibility is a concept by which the listener identifies an acoustic signal in the form of a voice code, somewhat in the manner that a card-sorting machine feeds cards to designated bins. Presumably the listener has prior knowledge of the code.
   b. Polytetrafluorothylene is a polymer that contains fluorine. It is produced by a reaction in a stainless steel autoclave which is similar to a large pressure cooker.
   c. Mantle convection is similar to the process of heating water in a pan, where the hot water rises to the surface, then cools by spreading outward, and finally sinks to the bottom to be reheated. The continents of the earth would split at the position where the hot water reached the surface. This point of upwelling is called a rise. They then would be carried away by the cooling water from the site of the upwelling to the position at which the cool water sank to be reheated. This position is called a trench. This, then, would be their final resting place until a new cell was created.
3. Discuss the effectiveness or ineffectiveness of the following, which the author identified as an extended definition.

Blue-Green Algae

Blue-green algae are plants which are high in protein. These plants can be used to alleviate protein malnutrition in this country and in undeveloped countries. The use of algae as a food supplement is not new. People in various parts of the world, particularly in the Far East, have used small portions of algae in their diets for centuries. There are various kinds of algae: green, blue-green, red, and brown. In general, a 100-gram portion of blue-green algae powder contains about 59 grams of protein, 19 grams of fat, 13 grams of carbohydrate, and 550 calories. Analytical data in terms of proteins, amino acids, certain vitamins, lipids, ash, and energy in various algae and mixtures of algae have been reported. The recent work of Vanderveen and his colleages suggests that some of the poor acceptability and lack of tolerance associated with algae as a food may be related to bac-

terial contamination. They have found that in "self-nourishing"
(autotrophic) algae cultures, as much as one-third of the cul-
ture population may be bacteria. However, work is now under way
to investigate the potential of decolorized algae as a food
source. The use of blue-green algae as food can solve the problems that confront a number of individuals.

CLASSIFICATION AND PARTITION

4. Classification is apparent in sentence 5 of "Blue-Green Algae," quoted in
   Exercise 3. What is the obvious basis for this classification? Consult your
   dictionary to determine if this classification of algae is the most meaningful
   that can be made.
5. Partition is apparent in sentence 6 of "Blue-Green Algae," quoted in Exercise 3. Is the partition complete and precise? Discuss.

DESCRIPTION

6. Discuss the effectiveness or ineffectiveness of the following device description.

                 The Electrostatic Precipitator

     The electrostatic precipitator collects particles by passing
a gas between two electrodes. These electrodes have a high-voltage,
unidirectional current flow.
     The electrodes are either collectors or dischargers. Discharge
electrodes are a series of thin wires. The collecting electrodes
have a surface 16 times as large as the discharge electrodes.
     Collecting electrodes are a series of parallel rods 0.1875
inches in diameter. They are suspended in a frame 1.5 inches
apart. These frames are 8.5 feet wide and 17.5 feet high. The
plates, spaced every 10 inches, form ducts through which the gas
passes.
     The discharge electrodes are in the center of the ducts. The
electrodes are 0.109-inch-diameter iron wire, spaced seven inches
apart. They are held taut at the bottom by the frame and a
weight.
     Hoppers at the bottom collect the precipitated particles. The
hoppers are built of heat-insulated steel. To prevent condensa-
tion on the hopper, a steam-heating coil is sometimes provided.
The whole construction of the precipitator is designed to prevent
corrosion.

7. Discuss the effectiveness or ineffectiveness of the following process description.

                 Sulfur Dioxide Conversion

     Processing sulfur dioxide in flue gases into sulfur is a mod-
ern concept of how to get rid of the worst industrial air pol-
lutant. Recently, several processes have been proposed for the

removal and recovery of sulfur from stack gases in commercially usable forms; but of all possible forms, elemental sulfur is most favored by marketing logistics.

The alkalized alumina process appears to offer the best prospects for commercial development. In this process a reactant (alkalized alumina) is subjected to alternating cycles of absorption and regeneration. Alkalized alumina is a physically active form of sodium aluminate. It is used in a fluidized bed absorber set at 300-350°C to react with sulfur oxides in the flue gases. After being heated to a temperature of about 650-700°C, the reactant containing sulfur is regenerated and recycled back into the absorber. Sulfur is removed from the solid.

SPECIAL PROBLEMS

8. Using your dictionary as a starting point, write a sentence definition of any five common hand tools, power tools, kitchen utensils, garden tools, bicycle (automobile, etc.) parts, sporting goods, etc. As your instructor directs, select items of one kind or of two or more kinds. Underline the term in each definition once, the genus twice, and the differentia three times.

9. Repeat Exercise 8 but select items whose terms must be qualified in some way. Remember to underline the complete term once in each case.

10. Repeat Exercise 8 but select items whose genuses (or genera as the plural is more often written) must be qualified in some way. Remember to underline the genus twice in each case.

11. Repeat Exercise 8 but select items whose differentiae must be qualified in some way. Remember to underline the complete differentia three times in each case.

12. Simplify the definition of a not-so-familiar item by using a synonym, an analogy, or a metaphor. Write one or more sentences, as necessary.

13. Write an extended definition of one item in one of the groupings in Exercise 8. Bracket and identify each of the individual orders of development you have used.

14. Present graphically the classification of two or more things of a kind that you are familiar with. Identify the basis of classification clearly.

15. Write a classification of two or more things of a kind that you are familiar with. Identify the basis of your classification clearly and include definitions.

16. Write a partition of an item that you are familiar with. Include definitions.

17. Write a description of a device that you are familiar with. Assume that your reader is unfamiliar with the device.

18. Write a description of a process that you are familiar with. Assume that your reader, an observer, is unfamiliar with the process.

19. Write a set of instructions for completing a commonplace act (tying a shoe, making a bed, balancing a checkbook, eating a lobster, etc.). Assume that your reader, a participant, wants to follow *your* instructions exactly.

Chapter Seven

SUMMARIES, LISTS, AND GRAPHIC AIDS

In industry today, a leader must know as much as he can about all conditions, developments, and problems in business generally, as well as in his own company. What he learns enables him to make good decisions. These decisions, in turn, help to ensure the successful operation of his company.

For most of his knowledge, the leader must rely on the reports of his workers. However, a majority of the reports have too many details for him to remember and keep everything in its proper perspective. Hence, he will request that his workers add summaries, include lists, and present graphic aids.

Summaries, lists, and graphic aids are simplifying and clarifying techniques in practical communication. A *summary* is the use of very few words to represent many. A *list* is the setting off of individual points so that they can be considered one at a time. A *graphic aid* is a visual presentation of facts and ideas in a table, chart, graph, etc. All three are highlighting techniques; that is, they are selected facts and ideas that stand out in reports. Therefore, all three contribute a great deal to clear communication.

SUMMARIES

Summaries are known by many names, such as abridgment, compendium, digest, précis, and synopsis. In technical writing, however, only two names are common—*abstract* and *summary* itself. Since these two are used with rather precise meanings by technical writers and editors, we shall discuss them separately.

Abstracts

Abstract is the name for the shortest complete summary. Two kinds of abstracts are written: topical and informative. The *topical abstract* is like an objective table of contents: it tells what a report or article is about but does not include findings (and conclusions and recommendations in interpretive studies). The *informative abstract*, on the other hand, tells both what a report or article is about *and* what it says.

Two examples clearly show the difference:

Topical Abstract

Two methods, solvent seasoning and vapor drying, are compared for drying wood by the application of hot chemical vapors. An evaluation is made on the basis of drying time, complexity of operation,cost, effect on strength, and creation of defects when wood of low moisture content is desired.

Informative Abstract

Two methods, solvent seasoning and vapor drying, are compared for drying wood by the application of hot chemical vapors. Solvent seasoning is found to be better when wood of low moisture content is desired. Solvent seasoning requires more time and is a more complex operation than vapor drying, and its chemicals are more expensive. However, it produces a higher profit because the wood is upgraded and can be sold at a higher price. Most important, it neither reduces strength nor creates as many defects as vapor drying, and therefore produces a more desirable product.

When you read the first example, the topical abstract, you learn only what the original report was about and what criteria were used for the evaluation. When you read the informative abstract, you learn everything essential, including which method is better and why.

Your Use of Abstracts

Both kinds of abstracts are useful, however. If you want to know only what a report or article is about, the topical abstract tells you enough. Then you can decide to read the whole report if the abstract arouses your interest in the subject. If you want to know everything essential, then what you want is an informative abstract. From it you may be able to learn what you need for your purposes. At least, you will know if in your own work you can probably use the findings—for example, for citations in a report you are writing.

Although topical abstracts are shorter, informative abstracts are preferred by most readers. In your own writing, therefore, you should include the informative kind. Leaders want to know "everything," of course. If they have confidence in you, they may read only your informative abstract, your lists of specific conclusions and recommendations, and your introduction. You may think that it is insulting for a leader to read only a small part of your writing. On reflection, however, you will realize that in doing so he is paying you a compliment.

Characteristics of Good Informative Abstracts

A good informative abstract has three major characteristics:

1. It is no more than 10% of the length of the original report or article.
2. It is faithful to the original in its coverage of important points, conclusions, and recommendations—to the point of including significant factual details.
3. It is faithful to the original in the handling of proportion and emphasis.

The importance of the second and the third characteristics is self-evident, but the first requires some discussion. Some readers insist on a length of less than 10% of the original. However, the number of words one should write depends on how much detail, and on how many conclusions and recommendations, the original contains. For example, 5% may be a realistic percentage for long reports but would hardly be adequate for short ones containing many details. For a 1,000-word report, 5% is only 50 words—three fairly short sentences!

How to Write a Short Summary (Informative Abstract)

(The heading says "Short Summary" because the technique can be used for all summary writing.) The characteristics of a good informative abstract indicate the requirements. But a step-by-step procedure for fulfilling these requirements is helpful to most writers. Such a procedure follows.

1. State in one sentence the central idea of the whole.
2. Jot down, in a word or phrase each, the important ideas that point to or support the central idea.
3. Establish tentative links (transitions) for putting these ideas into sentences.
4. Unless you are summarizing a process or a procedure, arrange the ideas in the order of their importance.
5. Establish permanent links to make sentences complete.
6. Add significant information that was left out because it was considered less important at first.
7. Write a complete draft, using the author's own words wherever possible, to give the summary some of the qualities of the original—and for the sake of accuracy.
8. Write other drafts, always trying to reduce the number of words and to simplify without altering the meaning, proportionate coverage, and emphases in the original.

By following these steps you should produce an effective, informative short summary.

Here are some other suggestions—both do's and don't's:

1. Read the original *several* times to become thoroughly familiar with it.
2. Estimate the number of words in the original. You can estimate easily by determining the average number of words per line and then multiplying this number by the total number of lines.

3. Note the maximum number of words you could use for a 10% summary. However, in your early drafts use more words, if necessary, to make sure that nothing important is left out. It is easier to cut down a summary's length than to fit words into a summary that is in "final" form. When you overwrite at the start, you also have a better sense of proportion and emphases.

4. Unless the topic sentences in the original are truly effective, do not merely copy them. Even if you can use the topic sentences, try to work in details. Remember that a good summary has *some* details. (Note, however, that minute details and specific examples can be included only in longer summaries—if at all.)

5. Try to reduce original sentences to clauses and phrases; then find transitions for linking these shorter structures to the others. In doing so, however, try to avoid writing new sentences that are cumbersome and too long.

6. If the summary must be fairly long, write several paragraphs, not just one. To save space, professional abstracters usually write single paragraphs. However, some one-paragraph abstracts violate unity and misplace emphasis.

7. Use Arabic numerals (1, 924) for numbers. However, convention rules that they never be used to begin a sentence.

8. Use abbreviations sparingly. It is expected that you will write *psi* for "pounds per square inch," for example. But words that are not abbreviated in original writing—such as *experiment*—should never be abbreviated in summaries.

9. Include all articles (like *a, an, the*) and prepositions (*of, by,* etc.). Phrases, clauses, and sentences must be fully developed in *all* kinds of writing.

10. When you finish, be sure to indicate clearly what it is that you accomplished. At the top of the finished draft include complete bibliographical information for the report or article (author's name, the title of his work, the name of the publication or the publisher if the work is a book; include volume and issue number for a journal article, and inclusive page numbers). Under this information put the word "Abstract," and under this, your name as its author. If you are abstracting your own original writing, omit the last, of course. If the abstract is to appear with the report or article you have written, only the word "Abstract" or "Summary" appears for identifying information.

Note that, characteristically, the final draft of an abstract is single-spaced.

Other Summaries

We shall discuss here three kinds of summaries that are identified by this name: introductory summary, text summary, and terminal summary.

Introductory Summary

Like the informative abstract, the introductory summary usually is placed just before the introduction of the report or article. In this position, the introductory summary is complete; it covers the whole of the original. If the introductory summary is placed after the introduction, however, information from the introduction is not included. Presumably, the reader will read the complete introduction just before he reads the summary. In either case, the summary conventionally is identified by the single word "summary."

Introductory summaries are longer than informative abstracts, running from 10 to 20% of the length of the original. Obviously, the longer summary may be needed when the original contains many important facts, conclusions, and recommendations. But if leaders prefer longer summaries for *all* reports, introductory summaries rather than abstracts should be written. Notice how much more detail can be included in an introductory summary of the same report as the informative abstract on page 122.

Summary

   This study, comparing solvent seasoning and vapor drying as the chemical methods for drying wood, shows that solvent seasoning is better when wood of low moisture content is desired.
   Both methods involve the use of heavy molecular organic vapors that are excellent conductors of heat for drying. The solvent-seasoning process uses acetone; the vapor-drying process, xylene. The major difference, however, is that solvent seasoning is a more complex operation and therefore requires more time. In the vapor-drying process, the water and the xylene are easily separated by mechanical means. In the solvent-seasoning process, on the other hand, extensive, time-consuming distillation is needed.
   For this reason and because the chemicals used in solvent seasoning cost considerably more, this method would appear to be less profitable. However, like vapor drying, solvent seasoning yields by-products (resins, and extractives for pharmaceuticals) whose sale results in profits that pay two-thirds of the operating costs. Furthermore, solvent seasoning produces a higher grade wood, and thus affords a higher profit. In one study, on the solvent seasoning of Ponderosa Pine, a profit of $4.30 per thousand board feet was realized before the wood was sold to lumberyards.
   Most important are the facts that solvent seasoning creates fewer defects, and that there is no loss in strength caused by excessively high temperatures. These higher temperatures, which are needed in vapor drying, cause the thermo-decomposition of lignin and cellulose, and therefore weaken the wood greatly. They also cause more checks and splits in the wood than are created by the use of solvent seasoning.

An introductory summary like this will please leaders because it tells them what, essentially, they need to know to make a decision. But it says little more than a good informative abstract if one overlooks the details. Thus we can say that, depending on what readers want, there is a place for both kinds of summaries in technical writing. Indeed, some leaders insist that long reports include both kinds.

Text Summary

As the name indicates, the text summary appears in the body of the report or article. Almost always it is quite brief—perhaps only a sentence or two. Ideally, it serves both as a review of the section just finished and as a transition to the next section. The following example, from the report for which the abstracts and the introductory summary specimen were written, illustrates this dual function:

    Because of the amount of distillation required, solvent
    seasoning is a more complex operation than vapor drying, and
    therefore requires more time. When these drawbacks are con-
    sidered along with the higher cost of the chemical used, sol-
    vent seasoning would seem to be impractical.
        However, ...

As the "However" suggests, a text summary is often used for emphasis—possibly, as it is here, to establish a sharp contrast. Obviously, a text summary will also be helpful after a long or involved section of information or discussion. Like all other summaries, this one is always prepared for the convenience of the reader. Of course, if emphasis is not needed and the development of the report is clear-cut, the text summary can be omitted.

Summaries of data (for example, in tables, charts, and graphs that are presented in the text), summaries of test results or of responses to questionnaire surveys, etc., also are included in the report or article itself. These summaries serve in the same way as text summaries: to simplify and clarify material for the reader's benefit.

Terminal Summary

A terminal summary, at the end of the report or article, is not always needed. A good abstract or a good introductory summary should be enough in many situations—especially the longer introductory summary. At times, though, a terminal summary is in order. For example, one should be included when the report is extremely long and full of details, and only an informative abstract has been presented. Also, a terminal summary should be included when the subject is controversial, and the writer thinks that a terminal summary will help to convince the reader of the soundness of the conclusions.

A terminal summary in some reports is a list of the important conclusions, numbered and indented from the regular left margin of the text. The conclusions may in some cases be listed in the order in which they appeared in the pre-ceding text. Usually, however, they are presented in the order of their decreasing importance. Such an order is both logical and necessary when recommendations follow. Recommendations should always be written in that order—the most important, first; the least important, last. The writer will have a crystal-clear ending if the order in both these lists is exactly the same.

The following lists of conclusions and recommendations, from a report en-titled "Estimated Digestibility of Crown Vetch in Beef Cattle," illustrates:

Summary of Conclusions

The major conclusions established in this report are these:
1. The Carroll variety of crown vetch in the first and second
   cuttings was more digestible than the Harrison and Williamson varieties.
2. Potassium and phosphorous fertilizers had little, if any,
   effect on making these three varieties more digestible.
3. The total nitrogen content in all varieties was greater in
   the second cutting of the plants.

Recommendations

On the basis of these conclusions, it is suggested that
1. The Carroll variety be used as the feed for beef cattle,
   in preference to the Harrison and Williamson varieties.
2. Potassium and phosphorous fertilizers not be used.
3. The second cutting be preferred to the first cutting.

No matter how it is presented, the terminal summary must remind the reader of the important points in the report or article. A good terminal summary brings into focus the points to be remembered for use later. And if a decision must be made now, it brings into focus the major conclusions and recommendations that your reader needs. You may, without meaning to, forget that a terminal sum-mary is a summary—and nothing more. Some writers try to introduce new ma-terial in their terminal summaries. But in a true summary nothing is included that has not been covered in the text of the report.

LISTS

Numbered lists are very helpful to the reader, especially when items are im-portant and the writer wants to be sure that no item is overlooked. If such items are presented in a regular paragraph development, the reader may have dif-ficulty in remembering each point. He may have even more difficulty in remem-bering which items are more important. Like summaries and graphic aids, lists are easy to remember because they are set off from the text. The examples pre-sented under "Terminal Summary" clearly show their virtues.

The examples also illustrate an important point about presenting a list. *Grammatically, every item should be the same.* If, as in the examples cited, the subjunctive mood was used, it should be used throughout. If questions are called for in a list, every structure should be a question. If phrases that function as nouns—to name conditions, for example—are needed, all should be written in the same way. And so on.

Numbering the items in a list also aids the reader. If he wants to discuss a point with the writer, all he has to do is refer to "Item 6"—or "Recommendation 2." This reference system is especially convenient for him when he must write to the author. Of course, everyone involved has the same advantage—including the writer himself.

Following is one more example. This time, noun phrases are used, but again a structure is used to introduce the list. Note that the structure becomes a complete sentence when each item is added to it. Note also that the points added in items 2 and 3 are themselves complete sentences. The reason for this is that they are not part of the basic list. The list is labeled "Highlights" because it serves as a summary of important information in a worker's report on a chemical society's meeting that he attended:

```
HIGHLIGHTS

    Highlights of the meeting, from the standpoint of our own in-
terest, were
    1. New developments in trace analyses for arsenic, sulfur,
       and vanadium.
    2. A new technique for analytical hydrogenation to measure
       unsaturation. It promises to supplement the methods used
       to determine the bromine number.
    3. Development of a cyanogen-oxygen burner for flame photom-
       etry. The high temperature of the cyanogen-oxygen flame
       permits the determination of many metals that until now
       could not be determined by flame photometry.
    4. The description of a technique for determining the compo-
       sition of equilibrium mixtures resulting from the iso-
       merization of pure hydrocarbons.
```

It goes without saying, that a list should be used only when needed and helpful as in both of the cases cited. Too many lists may confuse the reader and certainly will make each one less emphatic, and therefore seem to be of comparatively little importance. *Anything* that is overdone impresses the reader in this way.

GRAPHIC AIDS

Data in general and statistics in particular are found in all technical writing. When they are few in number and do not point to generalizations and conclusions, they can be presented in sentences and paragraphs—like any other information. However, when data are numerous, or point to important conclusions, the writer should use tables, charts, graphs, and other visual aids. "A picture is worth a thousand words" is more meaningful today than ever before. We live in a world full of statistics, and readers must see them clearly and in perspective to get the most out of them.

The purpose of this section is to help you see the value of graphic aids, to show you which are used for specific purposes, and to point out the requirements and the pitfalls that must be considered when they are used.

Usefulness of Graphic Aids

Statistics are used to show trends, comparisons, contrasts, results of tests, cause-and-effect relationships, etc. Obviously, when presented in graphic aids, such data help to emphasize your interpretation of them. Graphic aids are like summaries and lists in that they stand out on the pages of reports and articles. And like summaries, they compact information; that is, they present much in little. Indeed, many words would be required to present what a visual aid reveals in a brief examination. Graphic aids do not entirely *replace* words, of course; they do, however, enable you to use fewer words.

Graphic Aids in Common Use

The following review, with examples of graphic aids commonly used, will help you to see the potential value of visual presentations in your writing.

Table

A table is a grouping together of related numbers into either vertical or horizontal columns, so that the values can be easily referred to or compared. As the following example shows, a table should contain headings, and, like any other carefully presented list, it should use similar structures for all values of a kind.

Note the logic of organization in the table on page 130. The title covers all the lists; values and names are carefully aligned; and column headings are all clear-cut, as the underscores indicate. Even the order of listing the equipment is logical—from the largest number to the lowest.

Tables can be simple, like the illustration above, or quite complex, like the presentation on page 200. In all cases, however, they should be neat and symmetrical.

Line Chart

The line chart, also called the curve chart, is useful when you want to show either (1) the relationship of values, as in Figure 1 or (2) constant change—or no change at all—in values over a period of time. For three examples of change over a period of time, see pages 202-04.

Table 1

NUMBER, TYPES, AND CONSUMPTION OF UNITS OF AIR—OPERATED
EQUIPMENT IN THE OXIDE DEPARTMENT, JULY 1, 19__

Number of Units	Description	Present 100# Air Consumption	
		Maximum (CFM)	Normal (CFM)
67	Vibrator	980	365
5	Valve Packer	33	20
2	Bag Press	70	35
1	Air Motor Hoist	35	0
	Total	1,118	420

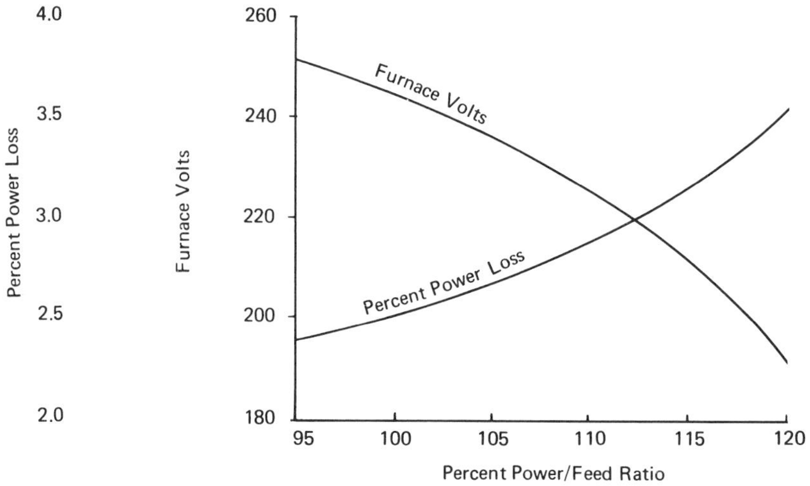

FURNACE VOLTS AND CORRESPONDING POWER LOSS
VS. PERCENT POWER/FEED RATIO

Figure 1

Bar Chart

With two exceptions, a bar chart also can be used to show relationships of values or changes in values over a period of time. Figure 2 illustrates the use of a bar chart to show relationships of values.

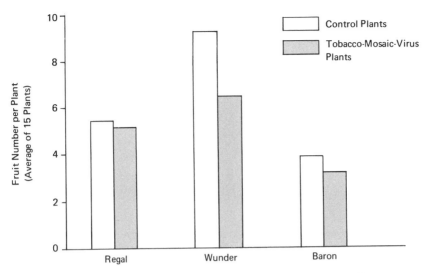

NUMBER OF FRUITS PRODUCED BY CONTROL AND TOBACCO-MOSAIC-
VIRUS PLANTS OF THE REGAL, WUNDER, AND BARON VARIETIES

Figure 2

The two exceptions are these: (1) If the base line represents zero, a bar cannot
be used to show a zero value, because a bar cannot be used to stand for zero. A
line chart, with a curve moving up or down, *can* show zero value as the curve
moves from or to the zero base line. (2) Obviously, a bar chart cannot show *continuous* change in a value—as a line chart can.

Segmented Bar Chart

The segmented bar chart, instead of a table, can be used to show the components of a value. Two or more segmented bar charts can be used either to show
differences in the make-up of a value at two or more times or to compare the
make-up of two different values at the same time.

Figure 3 shows the chemical components of a latex wall paint when it was
first manufactured in 1960 and when it was analyzed again in 1970.

Surface Chart

At times you may need to use a figure that is partly a line chart and partly a
segmented bar chart. Presumably, you would use the combination to show
either continuous change in the components of one large value or distinguish-
able variations in a single structure. An example of the first use would be a surface chart depicting the breakdown of total plant-operating costs, by month,
over the period of a year. An example of the second use is illustrated by Figure
8, page 137—a facsimile of a photomicrograph showing the phases, or separa-

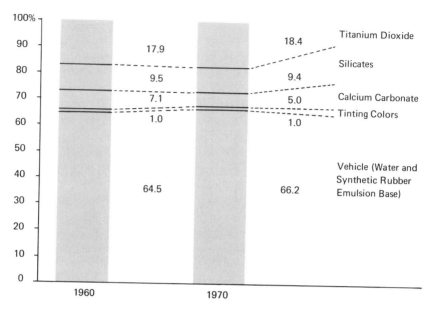

CHEMICAL COMPOSITION, IN PERCENT, OF LA-FLO PAINT,
1960 AND 1970

Figure 3

tions, in a galvanized coating. So that you can see how the facsimile explains the photomicrograph, both appear together on that page.

Flow Chart

The flow chart is used when you wish to illustrate either a process or a part of a process. Figure 4 shows a complete process: the sequence of steps and accomplishments required to prepare technical instruction manuals for customers' maintenance workers.

Map

A map is the figure to use when you want to show geographic areas. Values representing resources and economic growth are often shown in a map. To distinguish values, you can use different colors, shades of the same color, and various combinations of lines, hatchings, and dots.

Another common use of a map is for showing a plot plan. See, for example, Figure 6, which originally appeared in a report as a meaningful supplement to a photograph of the same area.

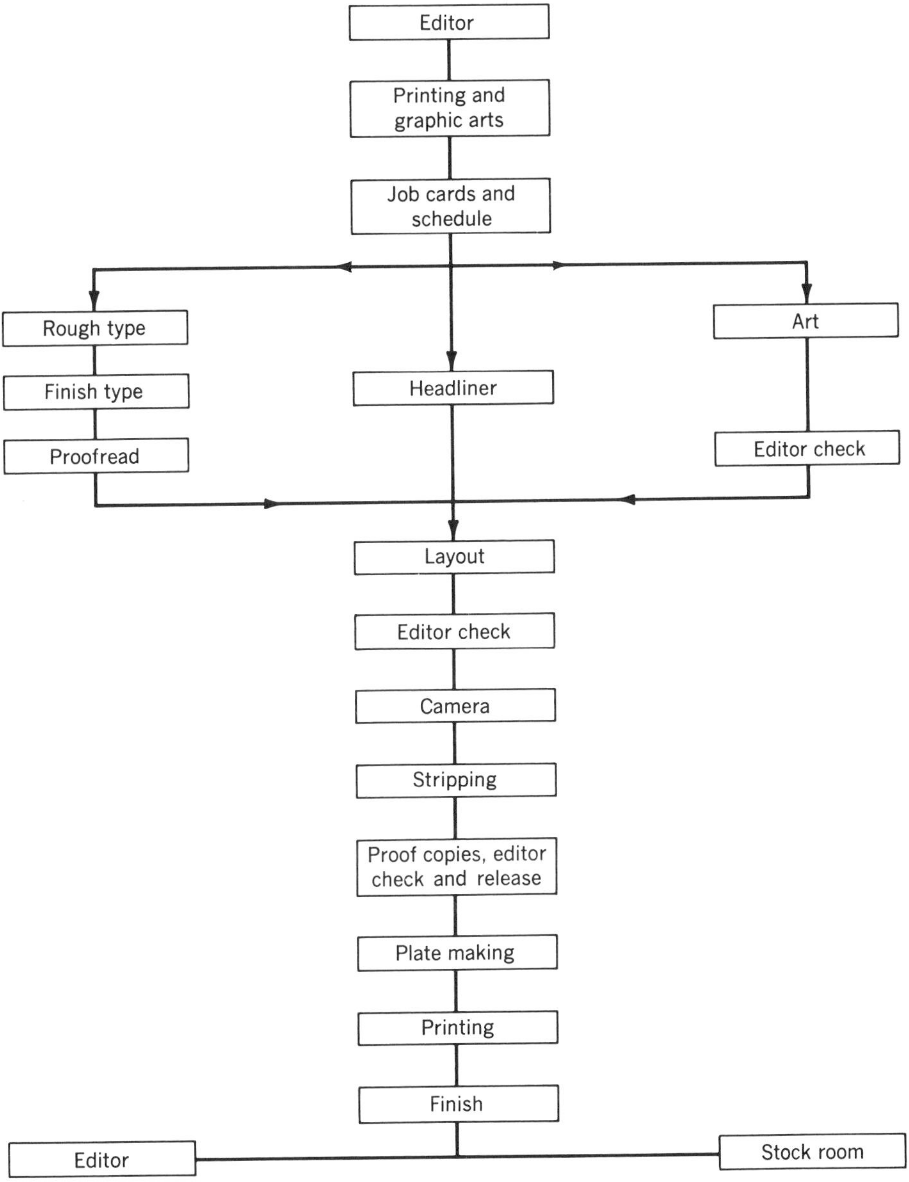

PUBLICATION PRINTING FLOW CHART

Figure 4

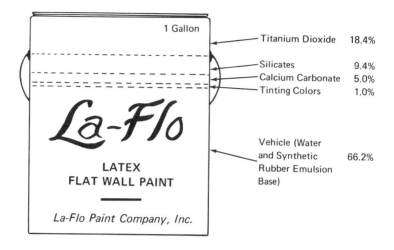

CHEMICAL COMPOSITION, IN PERCENT, OF LA-FLO PAINT, 1970

Figure 5

Pictograph

Although not commonly used in technical reports, the pictograph is very appealing to the eye. Actually, another way to present the data that usually are presented in tables or charts, the pictograph indicates breakdown, change, or distribution.

The pictograph of Figure 5 shows the information originally presented in a segmented bar chart on the right side of Figure 3.

Photograph and/or Line Drawing

The photograph and/or line drawing will probably be greatly appreciated by the reader of an industrial-data report on facilities, sites, etc., or of a technical report or manual containing precise details of what to look for during an inspection, disassembly, reassembly, etc. Either or both are also helpful to readers who are to follow instructions to carry out an operation.

Two illustrations of both together are provided. The photograph presented as Figure 7 is from an industrial-data report by the Greater Uniontown Industrial Fund. It appears with a map of the same area, Figure 6. The second example, Figure 8, is shown with its line-drawing complement, Figure 9. Having Figure 9

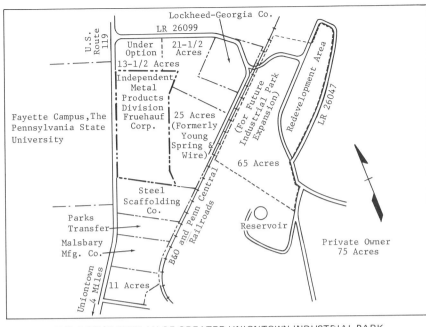

BUILDING/PLOT PLAN OF GREATER UNIONTOWN INDUSTRIAL PARK

Figure 6

Figure 7

AERIAL PHOTOGRAPH OF GREATER UNIONTOWN INDUSTRIAL PARK

on the same page also helps the reader to understand the photograph, a photomicrograph showing the phases, or separations, discernible in a microscopic view of galvanized coating. For another, separate drawing, see page 111.

Points to Remember in Preparing the Illustration

No matter which graphic aid you plan to use to emphasize a point, be sure that it is

1. simple,
2. accurate,
3. properly labeled, and
4. integrated with your text.

Let's look at each of these requisites, in turn.

Keep It Simple

You will notice that, with few exceptions, the illustrations are very simple; that is, only a few values are represented. Generally, you should keep the number of values in your illustrations low, also. If you must present complex tables, charts, drawings, and photographs in the text of your report, try to find ways of simplifying. For example, you can use color, clearly distinguished lines, contrasting surface areas, hatchings, etc., to help your reader see what you are attempting to point out. So long as you tell the reader what you are doing, you can even underscore important values in tables—or use arrows and other means of showing key values in charts and photographs.

Make It Accurate

If you distort values in presenting graphic aids, you are hurting yourself as well as the reader. A reader usually can tell when an illustration is inaccurate; when he can, he surely will have little confidence in the report—and in the writer in the future.

To be sure, some illustrations that appeal to the eye, like the pictograph, seem to indicate that a little distortion is all right. Indeed, it is difficult to make pictographs as accurate as tables and charts. But you should never deliberately distort. Notice what the pictograph in Figure 10 seems to show. The writer is graphically presenting the increase in a company's service manpower over a five-year period. As the white lines indicate, the service force has doubled in that time; the magnification of the representative worker's body, however, suggests that it has increased many more times than the actual figures *say*!

Among other ways to deliberately distort are these:

1. Disproportionately increasing the space between unit values on the vertical axis of a line or bar chart, while keeping the values on the horizontal axis the same, to make either the line steeper or the bar taller.

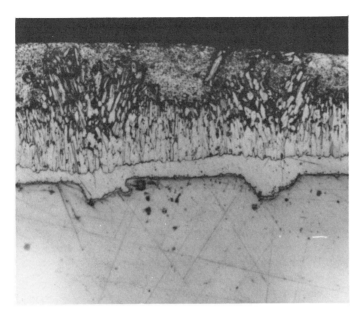

COATING ON SOUTH KETTLE STANDARD
SHOWING EFFECT OF STEEL-BASE SURFACE FLAWS
(Magnified 400X)

Figure 8

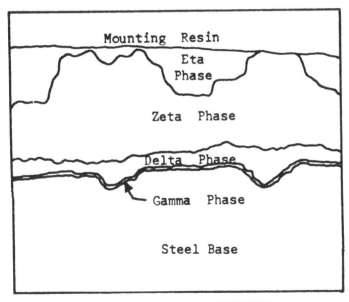

KEY FOR AREAS SHOWN IN FIGURE 8
(Magnified 400X)

Figure 9

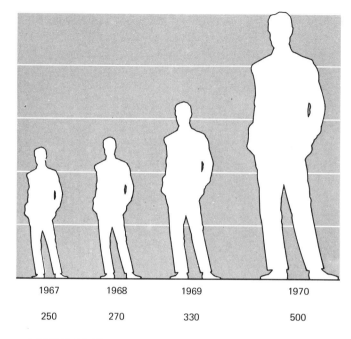

GROWTH OF SERVICE FORCE, MORROW COMPANY, 1967-1971

Figure 10

2. Drawing bars of different widths, so that one bar of a chart looks as if it represents a greater value than another bar.
3. Cutting off bar charts, that is, leaving out the bottom part of the vertical scale, so that all bars above the shortest look much longer than they are proportionately.

Label It Properly

Although graphic aids differ in appearance, all should be labeled properly with the following:

1. figure number,
2. title,
3. clear identification of values,
4. source note.

Figure number You should plan to assign a number to every graphic aid. Then you can refer to a particular illustration easily; so can your reader if he wishes to discuss it with you or anyone else. Of course, you may think it is illogical to present a very simple table or chart so formally. You will feel this way especially when you want the reader to hurry on to the text following. But when you must include several graphic aids, you will have trouble referring to particular ones, especially if you present more than one on a page.

Furthermore, if your report is long enough to contain a table of contents, you can list the graphic aids there by figure number, title, and page, and your reader can refer easily to any one he wishes.

In technical writing, two words, *table* and *figure*, are used to identify illustrations. Of course, "figure" covers *every* kind of graphic aid except a table.

Title   Every illustration with a number should also have a title. By including an appropriate title, you help the reader to clearly understand what the table or figure covers. "Covers" is an important word: it means that every aspect of the illustration is accounted for in the title. This point will be clear to you if you review the titles of the examples on the preceding pages.

If you don't use a title, very likely your references to illustrations will be vague and perhaps even misleading. In a table of contents, titles would obviously be more meaningful than numbers alone.

Clear Identification of Values   You should label every column, axis, line, bar, etc., so that the reader will know exactly what value each represents. As the more complex graphic aids show (for example, Figure 2 on page 131 and Figure 3 on page 132), the more values there are, the more you need to identify them precisely.

You should note also the use of special devices: the legend (explanation of values) in Figure 2 on page 131; the lines to the chemical components in Figure 3 on page 132; and the white border around the plot plan in the photograph (Figure 7) on page 135. By using these devices the writer makes sure that the reader will not be confused.

Source Note   Finally, whenever possible you should include a source note under every illustration you present. Such a note is the same as a footnote: it acknowledges and identifies the source of the table or figure. Even when you take data from a source and create an illustration of your own, you should show where the data came from. If the information came from several sources, use a footnote to tell your reader what sources were used. A beginning phrase such as "Data taken from ..." or "Information derived from ..." will enable you to use regular footnote form. (See pages 71–89.)

A source note should be used even when your table or figure is based on the results of a mail survey, on interviews, or on personal observation. For such sources, you can use your own words entirely. For example, you can say, "Personal Interviews: answers to Question 3," when you want to show by graphic means what your sample of respondents said on a certain subject.

The source note is included just as a footnote is on a page of text—at the bottom of the figure. (Most of the specimens are from confidential reports, and therefore source notes are not shown.) There are other ways of handling this kind of documentation, as the notes in the preceding two paragraphs indicate. But whatever technique you use (or are asked to use), you should always be as consistent as you are in footnoting.

Integrate It with Your Text

Although the point may be obvious, an illustration should never be used in place of words. In science, industry, business, and government, remember, writers always need to help decision makers as much as possible. Graphic aids can help greatly to satisfy readers' wants, and therefore you should use them freely. However, you should always make sure that readers understand why a graphic aid is being presented. Illustrations with clear and complete identifying information are only partly self-explanatory if the significance of their use is not also indicated in the text of your report. Thus, if a significant value in a table shows an improvement in the quality of finished products, you should highlight the fact in some way. For example:

```
As Table 3 shows, the quality of our finished products has im-
proved greatly.
```

A complete integration of a graphic aid with the text should include the following:
1. A summary statement of what the illustration's "message" is to the reader. (The excerpt just presented is an example.) This statement should include a conclusion whenever possible.
2. Specific facts that support the conclusion expressed in the summary statement. Such facts "drive home" the point so that the reader cannot miss it.
3. Further interpretation if more "meaning" statements can be drawn from the data of the table, chart, etc. When an aspect of a problem (or a solution) is being analyzed, for example, the writer should discuss any causes—reasons—that are indicated (and can be verified, naturally).

The last note can be related to the ideas covered on pages 83–93 in Chapter 5. The spirit of scientific interpretation should guide you in looking for explanations in the data that you have accumulated and plan to report. Also, however, the spirit of imagination should be allowed to range freely. If you continually ask the question, "What possible explanation can there be for this figure?" some meaningful facts, interpretations, and speculations should suggest themselves. When these are verified and they help to strengthen the message of your graphic aid, you should of course include them.

It may even be that hypotheses and ideas should be advanced when neither a definite conclusion nor an evaluation can be stated. If you phrase your speculations as such, your reader cannot say that you have drawn unwarranted conclusions. Such phrasing is a qualified beginning such as, "These data seem to show ... ." Speculations can be very helpful to your reader when strong conclusions or evaluations are unrealistic. As the saying goes, "Half a loaf is better than none." Naturally, no hypothesis or idea should be advanced that does not have at least *some* factual evidence to support it.

If you have not yet read Chapter 5, a study of the first few pages will help you to understand clearly the point of the final remarks in this chapter.

EXERCISES

SUMMARIES

1. Following are two abstracts of the same report, both composed by the original author. What weaknesses do you find in the abstracts? Does one do its job better than the other? Explain your answer.

Topical Abstract of "A Comparison of Two Common Detectors for Use in Gas Chromatography"

```
    A brief explanation is given on the thermistor detector
(a thermal-conductivity detector), the flame-ionization de-
tector (an ionization detector), and the fundamental process
involved in each detector.
    Detector requirements are named and briefly explained.
Each detector is evaluated by these criteria: sensitivity,
speed of response, linearity of response, stability, and ap-
propriate signal.
    Each detector is shown to be suitable for detection, al-
though a final judgment is made in favor of one.
```

Informative Abstract of "A Qualitative Comparison of Two Common Detectors for Use in Gas Chromatography"

```
    For many years the thermal-conductivity cells (katharom-
eters) have been the most widely used detectors in gas chro-
matography. But now new detectors such as the flame-
ionization detector are being employed.
    The katharometer is sensitive to all components, but the
flame-ionization detector does not respond to light gases or
water. The katharometer is somewhat slow in response but is
linear over a wide range. The flame-ionization detector is
not always linear in response but has a very fast response.
Each detector has stability problems. To amplify the signals
effectively, the katharometer needs only a simple electronic
amplifier, but the flame-ionization detector needs elaborate
electronic equipment.
```

2. Compare the following—the outline and the summary of a report on removing sulfur dioxide from a plant's flue gases. What weaknesses do you find in the summary? Explain your answers fully.

   a. **The outline**

```
Summary
Introduction
Processes for Removing Sulfur Dioxide
     The Reinluft Process
     The Catalytic Oxidation Process
     The Alkalized Alumina Process
```

```
Criteria for Technical Evaluations
     For the Reinluft Process
     For the Catalytic Oxidation Process
     For the Alkalized Alumina Process
     In General
Capital Investment
Operating Cost
Other Costs
Summary of Conclusions
Recommendations
```

b. The summary

Sulfur dioxide is one of the most commonly occurring contaminants of the atmosphere. An active interest in gas-cleaning
processes for removing sulfur dioxide from combustion gases has
been growing.

Earlier efforts in removing sulfur dioxide in low concentrations employed aqueous basic solutions. The removal of sulfur dioxide by these basic solutions worked well for some
processes. However, with sulfur dioxide being released from
burning fuels in greater concentrations, better processes for
removing it are needed.

Three processes for removing sulfur dioxide--the Reinluft
Process, the Catalytic Oxidation Process, and the Alkalized
Alumina Process--are furthest along in development. These
three processes have been developed to the stage where operating data and cost information are available and preliminary
estimates for installations can be made.

LISTS

3. The report that follows covers changes made in the specifications for a number of a company's products. Is the information presented effectively?
   Could it be presented more effectively in lists? Why or why not? Discuss
   your answers.

Attachments A, B, and C are revised sheets No. 4, 7, and
9 of Product Specifications.

Attachment A, sheet No. 4, replaces the sheet dated May 1.
On this sheet we have added the new brand, No. 6 Fuel Oil USL.
Corrections have been made on Heavy Marine Diesel Fuel and
No. 6 Fuel Oil CJ to conform to the current specifications for
these products.

Attachment B, sheet No. 7, replaces the sheet dated May 1.
We have revised the formulas for Motor Oil 10W and the Ex-
cello Motor Oil Series, except the 20/20W grade, to conform
to the current specifications for these products. The Motor
Oil H.D. Series, except the 50 grade and the Motor Oil X.H.D.
Series, have been revised to conform to the current specifica-
tions for these products. A new specifications sheet for
Motor Oil H.D. 50 was approved January 1, but printing of the

sheet has been delayed until the new Government "Qualified Products" number is received. A revision of Motor Oil X.H.D. 30 is being prepared now to adjust the viscosity at 210 F to approximate the upper limit. A specifications sheet for this product will be ready soon.

Attachment C, sheet No. 9, replaces the sheet dated May 1. We have added the new brands Special No. 153M Oil and Light Oil No. 7. We have revised F.S. Motor Oil 40, Light Oil No. 6, and No. 260 Motor Oil Series to conform to the current specifications for these products.

4. The following paragraph is taken from a report on improving the reproductive efficiency of mares. Is the information presented effectively? Could it be presented more effectively in a list? Why or why not?

Research efforts that offer the greatest promise of improving reproductive efficiency are those involving the use of such new techniques as superfetation, causing a pregnant female to carry two fetuses, of different ages, at the same time; superovulation, making an animal which usually discharges only one ovum at a time discharge a group of ovums at the same time; ovum transplantation, taking an already-fertilized ovum out of its mother and putting it in a host mother for her to bear and to care for after birth; freezing ovums, taking either fertilized or non-fertilized ovums and freezing them so that they can be stored and used at a later time; and estrus synchronization, artificially causing all the mares in a group to come into heat at the same time, so that it is easier to breed them.

5. Identify and eliminate any violations of parallelism in the following lists.

   a. The recommendations for a report entitled "Eliminating the Iron Contamination in the Zinc Plating of Sheet Steel:"

Recommendations

On the basis of preceding conclusions, it is recommended that:

1. The sheet steel be etched with a mild solution of acetic acid (10 parts acetic acid to 50 parts water), and then spray coat the steel with heavy-grade oil such as 50 weight.
2. Reduce the voltage to the electrode from 120 volts to 95 volts.
3. We add manganese to the electroplating solution in the amount of 4 grams per liter of the filtered solution.

   b. Excerpt from the body of a report entitled "Decreasing the Air Pollution at the Acme Pulp and Paper Company:"

The advantages of using this system are as follows:
1. No hydrogen sulfide is formed in the evaporator.
2. The forming of sodium hydroxide to help maintain a high pH.
3. There is the advantage of the formation of dimethyl disulfide, which is less volatile and much less objectionable.
4. There is the occurrence of less corrosion in the multiple-effect evaporator.

GRAPHIC AIDS

6. The following excerpt (including the table) is from the section "Concrete" in a report entitled "The Selection of a Floor for Hagen Chemical's New Plant." Does the table contribute to the effectiveness of the section? Why or why not?

Concrete

Concrete will not meet many of the standards for a floor in a corrosive environment. However, it is adequate for many areas, including warehouses, which make up much of the square footage in many plants.

Table 1. Characteristics of Concrete

Concrete	Chemical Resistance		Mechanical Resistance		Thermal Resistance	
	Acid	Base	Shock	Wear	Heat	Shock
Portland	*	***	***	*	**	**
Alumina	*	***	***	*	**	***
Phosphate bonded	***	*	***	***	**	**
Air-set silicate	*	**	**	**	**	**
Chemically set silicate	***	*	*	*	*	**

Although it may come as a surprise to people, all concrete is not the same. In general, there are two types of concrete floors: monolithic and granolithic. The other types of concrete are subdivisions of these two. The monolithic type is very sound structurally, but the surface does not withstand severe wear. The granolithic type, on the other hand, is essentially a surface finish. Its mix has a different water-cement ratio than that of the monolithic type.

7. Which of the illustrations in this chapter can be presented effectively in other (tabular or graphic) forms? Be prepared to discuss the effectiveness of the use of another form (other forms) in each case.

SPECIAL PROBLEMS

8. Write both a topical and an informative abstract of a writing assignment. Below each abstract record both the number of words in the complete original writing and the number of words in the abstract.

9. Write a complete introductory summary of a writing assignment. Below the summary record both the number of words in the complete original writing and the number of words in the introductory summary.

10. Write a text summary for inclusion in a writing assignment.

11. Review your last writing assignment to see where (if at all) you might have used one or more lists to advantage. Where it is appropriate, revise your writing to include the list(s).

12. Present one or more of the illustrations in this chapter in a different form (table or figure).

13. Interpret one or more of the illustrations in this chapter and write a report (reports) according to the plan recommended on page 400. You do not need to reproduce the illustration, but indicate where you would place it and leave an appropriate amount of space. (Naturally, the report you write will be limited to the data in the illustration in each interpretation.)

3

PRESENTING COMMON
FINISHED PRODUCTS

Chapter Eight

LETTERS

Characteristically, the letter has two features that set it apart as a form of communication. First, it is an *extra*company (rather than an intracompany) communication. Second, it can foster good public relations at the same time that it shares information. In other words, letters contain structures and words that clearly show the writer's interest in the reader. A good one-sentence definition is: *The letter is an extracompany communication which shares information while demonstrating empathy.*

We shall analyze first the layout—the physical appearance—of the letter. Then we shall examine the most common uses of the form in technical writing.

LAYOUT

Overall Layout

Since the letter is a personal representative of the company to outsiders, it should be laid out neatly and symmetrically.

A letter that is laid out symmetrically is balanced on the page. "Balanced" covers all of the common layouts: *block, semiblock, full-block,* and *simplified.* The most common layouts, block and semiblock, are illustrated on pages 158 and 159, respectively. In block layout paragraphs begin at the left margin, whereas in semiblock layout the first line of each paragraph is indented. The other two layouts, full-block and simplified, are illustrated on pages 162 and 163, respectively. Many organizations reject these layouts as too extreme. However, they are gaining in popularity because they save planning and typing time.

Margins

Since most business stationery is 8½ x 11 inches, we will use these measurements in setting our margins. If the size of the paper you are using is different, make proportionate adjustments.

All letters should have a top margin of at least one and a half inches and a bottom margin of two and a half inches. (In a letter several pages long, the bottom margin of each full page can be as little as an inch—as in this book's margin.) Side margins should be at least one and a half inches for a long letter, an inch and three-quarters for a medium-length letter, and two inches for a short letter. Table 1 shows precise margin settings for these three lengths.

Table 1

Left and Right Margins for Typed Letters

Number of words in body	SMALL TYPE		LARGE TYPE	
	Spaces in from left edge of page	Spaces in from right edge of page	Spaces in from left edge of page	Spaces in from right edge of page
Up to 75	25	22	20	17
76 to 150	22	19	17	14
151 to 225	19	16	14	11

The margins always should be set so that a rectangular frame of white space surrounds the typed letter.

Letter Parts

Every layout except the simplified layout has six parts: heading, inside address, salutation, body, complimentary close, and signature. The simplified layout has a heading instead of a salutation, and no complimentary close.

Heading

If you use printed letterhead stationery, begin typing at least a half inch below the bottom line of the letterhead. The heading on letterhead stationery consists of only the date, centered on the page in the block and the semiblock layouts; started at the left margin in the full-block and the simplified layouts.

If you use plain paper, begin typing about two and a half inches from the top for a short letter, two inches for a medium-length letter, and an inch and a half for a long letter. The heading on plain paper—called the *return address*—usually consists of three lines. The first line includes your street number and street; the second line, your city, state, and ZIP code; and the third, the month, day, and year. In the block and semiblock layouts, the heading is always to the right of the center of the page. In the full-block and simplified layouts, the heading always starts at the left margin set for the letter itself. For examples, see page 162 and page 163, respectively, for these positions.

Inside Address

After typing the last line of the heading, move down five or six spaces on letterhead stationery, two or three spaces on plain paper. At the same time, move the typewriter carriage to the right, so that you are ready to type at the left margin for the letter. When you have taken these two steps, you can begin the inside address—the name and address of the person or organization you are writing to.

Begin the first line of the inside address with the reader's courtesy or professional title when you write to an individual:

Courtesy titles: Mr., Miss, Mrs. (Ms. when you do not know a woman's marital status)
Professional titles: Dr., The Reverend, The Honorable, Professor, etc.

For a more complete list, consult the "forms of address" section of your dictionary.

Always address the reader by name, if you know it, and add his job title if you know it. When you do not know the name of the individual you want to reach, address the job title.

Examples: Mr. John W. Murray, Director of Research and Development
Director of Research and Development

When you know neither the name nor the title of the person you want to reach, address the company by its legal name.

Example: Swanson Chemicals, Incorporated

If you know the reader's name and title, your inside address will have at least four lines; the second line will be the company name; the third, the company's number and street address; the fourth, the city, state, and ZIP code. Of course, the number of these lines will vary according to the amount of identifying information that you know. For several examples, see the complete letters laid out on pages 158, 159, 160, and 161.

Note that no punctuation is used at the end of the line in both the heading and the inside address. Note also that only a few of the titles are abbreviated and that the only other abbreviations in the illustrations are "Jr." for "Junior" and "Inc." for "Incorporated": only legal abbreviations should be used.

A final point about the inside address applies to typing the envelope. The Post Office is now using a machine, called the Optical Character Reader and abbreviated as OCR, that can "read" up to 43,000 addresses an hour. So that the machine can operate, however, writers must satisfy the following precise requirements in typing the inside address on the envelope:

1. All lines of the envelope address must begin at the same left margin, as is illustrated by the inside address on page 158.
2. The last line of the address must appear as follows: City, State or Territorial Abbreviation,[1] Zip Code.

   Note "State or Territorial Abbreviation." Instead of typing out the name of the state or territory, the writer must use capital-letter abbreviations that the machine can read. The list of authorized abbreviations is given in Table 2.

Table 2

Alabama	AL	Montana	MT
Alaska	AK	Nebraska	NE
Arizona	AZ	Nevada	NV
Arkansas	AR	New Hampshire	NH
California	CA	New Jersey	NJ
Colorado	CO	New Mexico	NM
Connecticut	CT	New York	NY
Delaware	DE	North Carolina	NC
District of Columbia	DC	North Dakota	ND
Florida	FL	Ohio	OH
Georgia	GA	Oklahoma	OK
Guam	GU	Oregon	OR
Hawaii	HI	Pennsylvania	PA
Idaho	ID	Puerto Rico	PR
Illinois	IL	Rhode Island	RI
Indiana	IN	South Carolina	SC
Iowa	IA	South Dakota	SD
Kansas	KS	Tennessee	TN
Kentucky	KY	Texas	TX
Louisiana	LA	Utah	UT
Maine	ME	Vermont	VT
Maryland	MD	Virginia	VA
Massachusetts	MA	Virgin Islands	VI
Michigan	MI	Washington	WA
Minnesota	MN	West Virginia	WV
Mississippi	MS	Wisconsin	WI
Missouri	MO	Wyoming	WY

3. Room, suite, or apartment numbers must be typed after the street address (example: 1771 Williams Street, Apartment 9).

---

[1] The use of the state or territorial abbreviation in the heading and the inside address, as well as in the envelope address, may be widely accepted in the future. For the sake of appearance, however, especially in important letters, it is recommended that in the letter itself you continue to write out all state and territorial names in full.

4. The street-address line must appear immediately above the City, State or Territorial Abbreviation, Zip Code line.
5. The complete envelope address must not take up more than six inches in width and five typed lines in height, and it must end at least one-half inch from the right edge and one-half inch from the bottom of the envelope. If the envelope has a window, the space for the address must be reduced by one-quarter inch on all sides.

Salutation

After typing the last line of the inside address, double space and move the carriage back to the same left margin. You are now ready to type the salutation, your greeting to the reader. The salutation appears in all layouts except the simplified letter, where it is replaced by a heading comparable to the subject line in other layouts. (See page 163.)

Note that the salutation always agrees with the first line of the inside address. If you address an individual by name, the salutation is personalized also.

Example: (Addressee line) Dr. James W. Bent
(Salutation)   Dear Dr. Bent:

If you address a job title, use "Dear Sir" if you know the reader is a man but do not know his name; "Dear Madam," if the reader is a woman.

If you address the company name, use "Gentlemen," preferably. ("Dear Sirs" is also used, but less commonly.) If you have occasion to write a firm of women only, use "Dear Ladies" for the salutation.

Note the colon after the salutation in the examples. This punctuation mark is a convention in business letter writing.

Body

After typing the salutation (heading in the simplified letter), double space and again align the carriage on the left margin. Indent only if you are using the semi-block layout; otherwise begin the body of the letter at the left margin.

Paragraph Appearance Observe the length of the paragraphs in the specimen letters in this chapter. Most paragraphs are short, you will notice. On the average, they are seldom more than five or six lines long—considerably less than report paragraphs, for example. The length is in keeping with the brevity of most letters.

Note also that no letter has only a single paragraph. Paragraph unity is just as important in letter writing as in any other kind of writing. Remember, too, that special empathy is needed in your letters. Readers are inclined to think that you are in a hurry to get done when you write one-paragraph letters.

Presentation of Lists When you present a list, be sure to set it off from your regular paragraphs. Also, as the inquiry on page 159 illustrates, use numbers for items. Then both you and your reader can refer to an individual item easily. For full-block and simplified layouts, use the same margin that you use for the body. For the other layouts, you may indent.

Heading for Extra Pages When your letter is more than a page long, include identifying information at the top of each additional page. The conventional heading in block and semiblock layouts is a single line, as in this example:

```
Mr. Ralph W. Hughes                 2                    May 26, 19__
```

The heading begins at the left margin set for the whole letter and ends at the right margin.

The conventional heading in full-block and simplified layouts also begins at the left margin. However, the items are typed on succeeding lines rather than on one line. This example illustrates:

```
Mr. Ralph W. Hughes
May 26, 19__
2
```

All headings for extra pages begin about an inch from the top of the page. The continuation of the body usually starts three spaces below the heading.

Complimentary Close

Note that a number of words and phrases are used for the complimentary close—or "goodby"—in letters. As the following list shows, the close to be used depends on your relationship—assumed or actual—with the reader. The list is arranged in the order of decreasing formality:

Close	*Relationship*
Respectfully yours,	Formal
Very truly yours, (Yours very truly,)	Formal or Informal
Sincerely yours,	Formal or Informal
Sincerely,	Informal
Cordially yours, (Cordially,)	Informal (writer is close personal or professional friend of reader)
Best wishes, (Best regards,)	Informal (writer may or may not know reader well but wants reader to feel the relationship is friendly)

The most common closes are "Sincerely yours," and "Very truly yours." "Sincerely yours," is considered a little more informal.

Observe that the close is set off by double-spacing from the last line of the body. Observe also that the first initial of the first word is capitalized and that a comma appears at the end.

No complimentary close is used in the simplified layout. The writer goes from the last line of the body directly to the signature.

Signature

If the practice in your company is to present the legal name first in the signature, your signatures will look something like this:

MILTON LABORATORIES

James R. Connors, Chemist
Research Department

On the other hand, if your company permits you to write on your own authority—a growing trend—your signatures will resemble this example:

Sincerely yours,

James R. Connors, Chemist
Research Department

For either of these two signatures, double-space from the complimentary close[2] and then leave four or five spaces for your signature.

Special Devices and Techniques

Attention Line

The attention line directs a letter to a specific person in an organization when the company name is used in the addressee line. The attention line is placed two spaces below the last line of the inside address and two spaces above the salutation. This example illustrates:

Benson Electronics Corporation
4271 Haynes Boulevard
Bayville, Wisconsin

---

[2] Does not appear in the simplified layout, of course.

```
Attention: Director of Personnel
```

```
Gentlemen:
```

Note that the salutation agrees with the first line of the inside address, not with the attention line. Generally, you should use an attention line only when you prefer not to address your letter to a job title.

Subject Line

The subject line announces the topic of a letter. Usually, it is reserved for routine messages only—as in the illustration below:

```
Subject: Our Order No. N-590
```

Conventionally, the subject line is typed either two spaces above or below the salutation, or on the same line as the salutation.

Without the word "Subject," this line is standard in simplified layout and is always typed in all capitals. See the specimen on page 163 for an example.

Dictator-Transcriber Initials

If your letter is typed by someone else, the initials of the transcriber will follow yours as these illustrations show:

```
WAD/ah       LBT:mm       DIB/SW       TRP:EE
```

In all letter forms, the initials are typed from the left margin, two spaces below the last line of the signature.

Enclosure or Attachment Notes

Often you will need to enclose or attach another letter, drawings, plans, etc., either as exhibits or as references for your reader. Logically, when you have more than one exhibit or reference, you will present them in the order in which they should be looked at. The three examples below illustrate:

```
(1) enclosure or encl.    (3)  atts. 2: 1. Sketch of Present Plan
(2) attachment or att.                   2. Sketch of Revised Plan
                                encls. 2: 1. Your Specifications
                                          2. Signed Agreement
```

Companies have policies on whether "enclosure" and "attachment" are to be written out or abbreviated. When both attachments and enclosures are included, the attachment note is always placed first and is separated from the enclosure note by double-spacing.

The note or notes are always begun at the left margin, two spaces below the signature (below the initials if they are included).

Carbon Copy

When you send copies of your letter to others, type out or abbreviate "carbon copy" ("cc") two spaces below the last typed element—as indicated by these examples:

```
cc: Dr.F. B. Hall            cc: Mr. O. F. Loomis
                                 Mr. W. N. Foley
```

"Copy to" ("Copies to") also may be used, but it is never abbreviated.

The "carbon copy" or "copy to" note is begun at the left margin.

The most common letters written by technical writers can be grouped into four classes: (1) requests and inquiries, and replies; (2) orders and invitations to bid; (3) claims and complaints, and replies; and (4) notices, announcements, and other information-giving letters.

REQUESTS AND INQUIRIES, AND REPLIES

Requests and Inquiries

The request letter and the inquiry letter are used to ask an outsider one or more questions. A simple request is at one extreme; a fairly long inquiry asking for specific information is at the other. We will examine these two types in their most common developments.

Simple Request

As the illustration on page 158 (in block layout) shows, a simple request is easy to write. It does, however, require a special concern for demonstrating empathy. This means that you should consider using more civilities (courteous or polite notes) than the mechanical "please." The following outline illustrates how a simple request can be developed effectively:

Paragraph 1. A courteously stated explanation (reasons) for writing.

Paragraph 2. The request itself, along with any further explanation that may be needed by the reader.

Paragraph 3. A courteous ending, indicating urgency if necessary.

Z E N I T H   L A B O R A T O R I E S

NEW YORK DIVISION
18 South Street
Manfred, New York 13466

June 12, 19__

Crescent Instrument Company
49 Porterfield Avenue
New York, New York 10091

Gentlemen:

We will sincerely appreciate your help in obtaining information
about low-range pyrometers for measuring surface temperatures
from 100 to 500°F.

Do you make one or more pyrometers for surface-temperature measurements? Will you please send us a copy of your latest catalogue or copies of catalogue sheets describing pyrometers that
you can make available?

Your prompt reply will be gratifying because we will need a
pyrometer within three weeks for an important phase of experimentation in a current project.

Sincerely yours,

ZENITH LABORATORIES

Ronald E. Hunter, Director

REH:lr

A Simple Request Letter Illustrating the Use of Block Layout

Multiple-Question Inquiry [3]

During research, you may need to write to an authority on the subject you are
investigating for up-to-date information.

When you know exactly what data you need, writing a multiple-question inquiry can be a simple task. However, you must explain fully why you are writ-

---

[3] For more detailed advice on asking questions, see pages 73–75.

ing. You must also be sure that every question is clear. And you must write in a way that will interest the authority in your problem and that will motivate him to answer promptly. The example below (in semiblock layout) will help you to see the logic of an outline like that on page 160.

19-B Crawford Hall
Eastern Junior College
Allegheny, Pennsylvania 15515
October 18, 19__

Mr. Frank W. Dibble, R-4
National Motors Corporation
1622 Commerce Boulevard
Commerce, Michigan 48597

Dear Mr. Dibble:

Your article, "Casting the Aluminum Engine," in the October issue of <u>Automotive Engineering</u> is the most interesting account of the Aleng casting process I have read. In fact, I was so impressed that I have elected to write on the subject for an informational-report assignment in English 20, Technical Report Writing, this fall.

Before I read your article, I had already seen several other articles on the subject. Your article has by far the largest amount of useful data, but I still lack the answers to several questions that will make my report complete:

1. Why is a cold chamber used when a hot chamber is said to produce better results?
2. How much, approximately, does the die cost?
3. How many blocks can be cast before the die has to be replaced?
4. What is the weight of the shot?
5. How many blocks can be cast per hour?

I shall very much appreciate receiving your answers to these questions, as well as any new data accumulated during your research since you wrote your article.

My report is not due until after the first of November, but an early reply will enable me to put your information in its proper perspective.

Sincerely yours,

John N. Bancroft

A Multiple-Question Inquiry Illustrating the Use of Semiblock Layout

Paragraph 1. A courteously stated explanation (reasons) for writing.

Paragraph 2. The request for information, made in a generalized statement that points to:

Paragraph 3. A list of *clear,* specific questions, numbered in sequence and set off from the rest of your letter.

Paragraph 4. An invitation to send any additional information that the authority thinks you will find helpful.

Paragraph 5. A courteous ending, indicating urgency if necessary.

This outline covers the essentials. Naturally, in a certain case a special development may be needed. For example, your explanation for writing may require several paragraphs. Or, you may have to discuss at length something that the authority has written or lectured on previously. Or, you may want to discuss points made in recent articles by other authorities.

Two other points about inquiries should be made. First, you may need to offer to restrict your use of the authority's information—it may be confidential. Second, you may need to be specific about your deadline for receiving the reply. Always allow at least a week to get the information into your report "in its proper perspective"—as the sample letter says.

Replies to Requests and Inquiries

When *you* receive a request or an inquiry, you can write three possible answers. You can comply, you can refuse, or you can write a reply that falls somewhere in between. The last two answers should be discussed together because for many people a "somewhere in between" answer is little better than a complete refusal.

"Yes" Replies

When you answer a request or inquiry positively, you have an easy task. However, you should realize that as much empathy may be needed in a positive reply as in a request or inquiry. It is not enough to "just do what you were asked," in a matter-of-fact way. Nor should you ever be grudging in either granting a request or supplying information. In addition to receiving what you have to supply, readers want to feel that you gave it to them gladly.

If you *are* reluctant to reply to a request or inquiry, perhaps you should do some rethinking before you send the letter. If you cannot honestly and courteously grant the request (as the illustration on page 162 does), perhaps your answer should be either a partial or even a complete refusal.

Note that the example on page 162 (in full-block layout) includes salesmanship as well as demonstrates empathy. Although use of the technique is especially appropriate in this kind of letter, occasionally it may be "right" in *your* letters, too. Whatever the nature of your specialized technical work, in your organization as a whole everyone's objective is to continue operation. As a member of the organization, you can sometimes make a contribution by using a little "salesmanship" yourself. Of course, your use of the technique must be sincere and honest, as it is in the letter cited.

"Partially Yes" and "No" Replies

Unfortunately, you cannot always write a completely positive answer to a request or inquiry. At times you must say "no" as part of your reply, and once in a while you must say "no" completely. In either situation, you should know that the reader is sure to be disappointed. To be sure, he will be grateful if most—even only some—of his request is fulfilled. However, the part of the request you cannot grant, or the information you cannot send, may in a particular case be *most* important to him.

Suppose, for example, that the writer of the letter on page 158 had previously written to *another* supplier, who was unable to supply a pyrometer for a month or more. Director Hunter would probably have been irritated because he would then have to send his request elsewhere. And you can imagine how poor handling by the other supplier might have ruined any chance for a business relationship with Zenith at *any* time.

Fortunately, there *is* a way of handling either a partial or a complete refusal without risking such irritation. From the time of Aristotle, man has known that a *reasonable explanation first* makes it easier to say "no" either partly or wholly. It also helps to make such an answer more acceptable to the reader.

The illustration on page 163 (in simplified layout) is an example of an effective "explanation first" letter. Note that characteristically such a letter follows this outline:

Paragraph 1. A courteous acknowledgment of the request (inquiry).

Paragraph 2. A complete explanation presented *before* the refusal.

Paragraph 3. An indication of an effort to demonstrate empathy somehow, if possible.

Paragraph 4. A courteous ending, perhaps with a little salesmanship and, if necessary, a statement or restatement of regret.

Where something positive can be said, paragraphs 2 and 3 should be altered. Paragraph 2 should state what *can* be done; paragraph 3, with an explanation first, what can*not* be done. You should always point up what you can do—and play down what you cannot do.

ORDERS AND INVITATIONS TO BID

Orders

To be effective, an order for instruments, machinery, etc., should anticipate a number of questions. After the introductory paragraph, the writer should specify throughout exactly what he wants. Answering the following questions will help to ensure that the supplier's informational needs are fulfilled.

1. How many? (number of units of the item)
2. What unit? (feet, rolls, dozen, gross, etc.)
3. What product name? (If the writer does not know the product name, he should identify the product precisely. Then the supplier will be able to

C R E S C E N T   I N S T R U M E N T   C O M P A N Y

49 Porterfield Avenue
New York, New York 10091

June 14, 19___

Mr. Ronald E. Hunter, Director
Zenith Laboratories
New York Division
18 South Street
Manfred, New York 13466

Dear Mr. Hunter:

We sincerely appreciate your letter of June 12, asking us about low-range pyrometers.

Enclosed is our latest catalogue, along with two supplemental data sheets on pyrometers we make that are used in measuring surface temperatures. The data sheets were prepared especially for you by our engineers. These supplements should help you to decide which of the two models described on page 16 of the catalogue is better for your experimental work.

As you may know, our pyrometers and other instruments have been highly rated by all standards bureaus and are guaranteed to be accurate for precise measurements. We take a great deal of pride in the recognition we have been given throughout industry.

Mr. Hunter, your order will be shipped the same day it is received here. We look forward to serving you both now and in the future.

Cordially,

Paul W. Smith, Manager
Sales and Service

PWS:ni

A Reply to a Request Letter Illustrating the Use of Full-Block Layout

S T A N D A R D I N S T R U M E N T S, I N C O R P O R A T E D

501 Winfield Street
Boston, Massachusetts 02220

June 9, 19__

Mr. Ronald E. Hunter, Director
Zenith Laboratories
New York Division
18 South Street
Manfred, New York 13466

YOUR LETTER ABOUT PYROMETERS

Thank you for your letter of June 6, seeking information on low-range pyrometers for measuring surface temperatures from 100 to 500°F.

We do indeed manufacture a pyrometer for just the application you have in mind. Like all the other instruments we make, our pyrometer has been available in any reasonable quantity desired. Three weeks ago, however, we received an extraordinarily large order from the government. We are therefore unable to accept your order for delivery within three weeks.

An adjustment in our manufacturing schedule will make it possible to ship pyrometers to other customers during the first week in July. On the assumption that you may be able to wait until then, we are enclosing our latest catalogue and price list. A complete description of our pyrometer is on page 24, and the price list is on page 30.

We think that our pyrometer, like the other instruments we make, is the best that can be obtained anywhere at home or abroad, Mr. Hunter. We should like very much to have a chance to prove our claim now or in the future. Please write again if we cannot serve you at this time.

Henry C. Wilson
President

HCW:MR

A Reply to a Request Letter Illustrating the Use of Simplified Layout

send what is wanted. Some companies have individual numbers for their products. Thus, the writer may have to use only that number and perhaps a brief, simple name for the product.)

4. What size? (If a product is manufactured in different sizes, the writer must be sure to spell out the one he wants. For example, he should write "small," "medium," or "large"; or, "1-oz.," "4-oz.," etc. Some suppliers number package sizes, so that by using a number the writer will be sure to get the size he wants.)

5. What price? (The writer must be careful to add the price if it varies from time to time. For instance, in some industries prices vary with the seasons. If the buyer wishes to pay no more than a certain price, he must state that price.)

The writer should also include the following information in the order:

1. If necessary, the means of transportation (rail freight, truck, Railway Express Agency (REA), parcel post, etc.) and the routing information to be followed. New suppliers are grateful for this information because purchasers know what means of transportation to their plant is the fastest and least expensive.

Many purchasers specify that delivery be made FOB (free on board)—and then add immediately the point or place on their property to which delivery is to be made. FOB means delivered to that place at the supplier's expense. Naturally, since shipping costs can mount up, making such an arrangement is highly advantageous to the purchaser. You should note, however, that some suppliers will permit such an arrangement only for certain kinds of goods.

2. Date of shipment desired. Some purchasers want items delivered no later than a specific time. Thus, they will say that a shipment will be accepted only if it is received by the date stated.

Other notes that are not usually included in orders are sometimes needed. For example, when goods are to be shipped elsewhere than to where the purchaser is located, the correct destination must be stated. Or, it may be that terms of payment, differing from usual terms, must be spelled out. Or, the signature of a high official in the purchasing company may be required in an extraordinary situation.

Because orders are considered routine and mechanical, very little extra empathy is needed in most cases. To suppliers, "getting the business" is often the only courtesy desired. However, as the following example shows, a little extra empathy goes a long way toward ensuring a pleasant continuing business relationship:

```
Gentlemen:

    You will be glad to know that you were the low bidder again,
in response to our request for delivered card prices. Will you
therefore enter our order for the following:
```

3,240,000 No. 534796 Voucher and Time Clock Cards, Manilla,
          upper left corner cut, per attached sample.
   50,000 No. 534797 Voucher and Time Clock Cards, Solid
          Salmon, upper right corner cut, per attached
          sample.
  100,000 No. 534706 Material Detail Cards, Manilla, Red
          Stripe, upper left corner cut, per attached sample.
Of course, this order is a commitment only at the prices
specified in your bid No. 29117-A, dated July 7, 19__, all de-
liveries to be made FOB our plant.
    As before, we will appreciate your delivering the cards ac-
cording to schedule, as follows:
    The Manilla Voucher and Time Clock Cards (No. 534796) are to
be delivered the first week of each month, in quantities of
270,000 per month, for twelve consecutive months beginning
August 1.
    All of the other two kinds of cards are to be included in the
first shipment of the Manilla Voucher and Time Clock Cards.
    We are pleased to be able to do business on a continuing
basis and will appreciate your prompt acknowledgment of our order
and delivery schedule.

                           Very truly yours,

Invitations to Bid

The invitation to a supplier to bid is no less precise than an order and also may
have to demonstrate a little extra empathy. To most suppliers, that "extra empathy" is gratifying when explanation is included with it. The following example
illustrates:

Gentlemen:

    We will appreciate your quoting in duplicate the price,
weight, point of shipment, terms of payment, and earliest delivery date, FOB our plant, of the tanks listed on Specification
Sheet No. 822-L, attached.
    To help you estimate, we have also attached drawings showing
where provision must be made for fittings in the fabrication of
the tanks. Please note particularly that fitting location E is
new for this type of tank.
    We have arranged for all bidders to inspect the erection site
at 10 A.M. on May 21, 19__. Will you please tell your representative to ask at the gate for Mr. N. L. Muncy, Project Engineer,
Engineering Design Department. Mr. Muncy will review the drawings
and answer any questions on this date only.
    Please include in your bid the date you will be able to de-
liver the tanks to the site, and specify the number of days you
will need to erect them.
    Before fabrication, the successful bidder will be required to
submit two sets of detailed drawings for our approval. Before

erection, he will have to submit six sets of prints bearing evidence of our approval.

Please note that May 29 is the deadline for all bids.

Sincerely yours,

CLAIMS AND COMPLAINTS, AND REPLIES

Problems arise in the normal operation of any organization, and many of these problems are created during relationships with other companies. An error is made in packing, shipping, or billing merchandise. A product either malfunctions or fails completely. A product or service is found to be less satisfactory than a guarantee or promise had indicated it would be.

For one of these reasons or another, a customer writes to point out an error or discrepancy. Then the organization receiving his letter must make a decision. The organization must do one of three things: (1) make a complete adjustment, (2) make only a partial adjustment, or (3) refuse to make any adjustment at all. In each case, the answer must be made acceptable to the customer.

Claims and Complaints

First, of course, the customer must write an effective claim or complaint. A claim or complaint must always be well substantiated. Details including dates, etc., must be spelled out; and the customer must show clearly that he is entitled to an adjustment.

A logical outline for a claim reflects this concern for being specific:

Paragraph 1. A courteously stated listing of the most significant details.

Paragraph 2. A presentation of all facts pointing to the claim itself.

Paragraph 3. A clear-cut statement of the adjustment desired.

Paragraph 4. A reasonable argument substantiating the request for adjustment and motivating the reader to comply.

Paragraph 5. A courteous ending, perhaps with the writer's offer to do something to obtain a prompt adjustment.

Following is an effective claim developed from such an outline:

Gentlemen:

Thank you for your invoice No. 41226, dated January 9, covering the shipment of one Model 111 transistorized A-Gauge Thickness Tester.

When this instrument arrived two days ago, our employees in the Receiving Department found it to be in the following condition:

1. The outer case could not be closed tightly.
2. All screws on the hinges and the locking device were loose.
3. The two halves of the outer case were not in alignment.

4. The locking device would not catch and hold.

5. The screws holding the mechanism were loose.

We should like to return the tester at once at your expense, and receive an acceptable replacement.

This is the first instrument we have purchased from you that was not in excellent condition. Clearly, the tester does not measure up to the quality we have found in all the other instruments that you have supplied.

Because we need an effective tester at once, we are returning the instrument by parcel post today rather than waiting for your authorization to ship. Please credit our account and bill us separately for the replacement. We will deduct the cost of return postage from the new invoice.

Complaint letters are similar but differ in that no adjustment is specified. In each case, the writer leaves it up to the supplier to decide what to do. Sometimes, of course, the writer expects nothing in return. He just wants the supplier to know that a complaint is in order.

In the following illustration, you will notice, nothing can be done about purchases made in the past. However, the writer makes it clear that the supplier will be in trouble if corrective action is not taken in the future.

Gentlemen:

On May 23 we received from you 1788 lb of AA Long Wool Yard and Horsehair mixture in response to our order No. FD-5619.

When we analyzed the mixture, we found that the fibers of animal origin in this shipment amounted to 58.3%. This figure is obviously well below the minimum specification of 94.0%.

We have been checking your shipments carefully during the past three months. In each case we have found a lower percent of fibers of animal origin; the figures are

Date Received	Order Number	Fibers of Animal Origin (%)
March 10	FD-5012	78.2
April 20	FD-5389	68.1
May 23	FD-5619	58.3

Our specifications for this material also require that the manufacturer supply a notarized statement with each shipment. We have not received this statement with any of the last three shipments.

So that these discrepancies will not occur again, we are sending a copy of our Specifications Sheet GT439, which we revised in January last year, sixteen months ago. The sheet clearly states the minimum specification of 94.0%.

Since we accepted your bid for AA Long Wool Yard and Horsehair mixture for one year, we shall continue to expect shipments from you according to our contract. However, we will be unable to approve payment in the future unless the notarized statement is received with each shipment.

We regret this alteration in what had been a very satisfactory business relationship. Perhaps the error in each of these three cases was inadvertent. We sincerely hope so, and are proceeding on the assumption that it will not happen again.

Very truly yours,

We cannot help marveling at the writer's (company's) patience and generosity in this letter. However, honest mistakes do happen. When no serious harm comes, the best course of action is to complain strongly but to give the supplier the benefit of the doubt. Obviously, too, it is *more* than unlikely that the error ever happened again: no reader could miss the implied threat of legal action!

Replies to Claims and Complaints

Just as in replies to requests and inquiries, explanation must dominate letters sent in response to claims and complaints. The one difference, naturally, is that explanation is needed when a complete adjustment is made, not only when either a wholly or a partially negative answer must be sent. Again, too, the explanation should precede the statement of the supplier's decision, except when an adjustment is granted or indicated—as in the situation evoking the complaint letter just quoted. Then, as is appropriate in the case cited, positive action should be stated immediately. The first two paragraphs of the reply to this complaint illustrate:

Dear Mr. Lyons:

You certainly can expect that the error referred to in your letter of May 27 will not happen again. You can also look for an extra 25% discount in your next order of AA Long Wool Yard and Horsehair mixture. This adjustment, we sincerely hope, will make amends for the errors made in our shipments to you. Truly, we are sorry.

As you may know, we prepare hundreds of mixtures of animal and synthetic fibers. Unfortunately, we have just learned, your last three orders were shipped to another customer. However, we have taken steps to ensure that 94.0% is the minimum amount of animal fibers you will receive from now on. And the notarized statement to attest will accompany each shipment, as specified.

We can see that it was well to give the supplier the benefit of the doubt in this case. The problem was resolved to the customer's satisfaction as soon as this reply to the complaint was received.

Two other points about replies to inquiries should be made. First, when a complete adjustment should be made, the writer should never grant the adjustment grudgingly. Second, when the customer has done something wrong to make a product malfunction or fail, and the claim is granted, he should be educated on how to use the product correctly. The instructions should be stated tactfully, of course; but before the grant is made, clear instructions should be presented to forestall another malfunction or failure.

NOTICES, ANNOUNCEMENTS, AND OTHER INFORMATION-GIVING LETTERS

During the normal course of a company's operation, occasions arise to write to customers, suppliers, etc., on a variety of subjects. It would be impossible to cover all of these in a book, much less in a chapter like this. A few illustrations of letters written for several common reasons will show you just a little bit of this broad range.

First is a letter to a possible supplier which indicates that further correspondence is needed before business can be transacted:

> Thank you for sending us a sample of your new additive, EN-211.
> After a number of experiments we have concluded that we can use this additive, with some minor changes, in our blending process. We are attaching a list of the changes that would make it possible for us to use your product.
> If you can make these changes, will you please tell us how much the additive will cost per gallon and when we could expect to receive your first delivery?
> If you have any questions about the changes we recommend, please feel free to call me at any time.

Next is a letter which responds to a request for carload shipping instructions:

> Thank you for your request of June 30 for carload shipping instructions to cover the movement of the Raschig Rings specified in our order BXX-53466.
> We have no particular preference concerning the main line to be used, although of course delivery to our plant must be made by Penn Central. If you are not on a Penn Central spur, we suggest that you switch to Penn Central in Akron.
> We assume that with this information you will be able to route the shipment without any difficulty.
> As agreed, shipment will not be made until we let you know when we can make arrangements for storing the rings. Indications are that we will be able to give you this information by mid-July.

Finally, we shall look at a letter that rejects a proposal and then makes an important request:

> Thank you for your letter of April 22 transmitting prints of an arrangement for a carriage track for removing coils on the reactors under construction at our plant.

We have carefully reviewed your drawings and sincerely appreciate the amount of work you have put into them. Unfortunately, our Maintenance Department will need a more permanent structure to handle the coils. Therefore, we must reject the arrangement you have proposed.

Instead, we are preparing our own drawings of a permanent structure that will meet all our Maintenance Department's requirements. The prints will be completed and mailed for your review by May 20.

As you know, we will furnish all of the material needed for the fabrication of the new structure. For this reason, and because erection of a permanent carriage track instead of a temporary one will be to your advantage, we should like to know if you will complete the job without increasing the contract price.

Will you please give us your answer to this question as soon as you have reviewed the new drawings?

TONE AND LETTER LANGUAGE

The last section of this chapter discusses two major points about the letter that are implicit in the definition but need to be amplified. These points are that the writer's tone and language can make a great deal of difference in the way his reader reacts.

Tone

Letters that demonstrate empathy are developed with positive tone. Of course, you now know that positive tone does not mean that the message is positive. Tone is the way in which the writer reveals his disposition and attitude toward the reader. Positive tone means that the writer looks at his message and the reader in an interested and friendly way. His message may be either wholly or partially negative, but his handling makes it as concerned and amicable as the situation permits.

Words themselves help to give a message its tone. Following is a partial list of "positive" words; these are words that all of us like to see in the messages we receive. After this list comes one that lists a few of the "negative" words in letters; these are words that all of us do *not* like to see.

Positive Words

appreciate	efficient	notable	service
benefit	excellent	please	sincerity
capable	faith	prestige	success
commendable	friendly	recommend	thank you
cooperate	grateful	reliable	thoughtful

Negative Words

allege	complain	impossible	state (verb)
apology	decline	liable	unfair
blame	disappointed	must	vague
careless	doubtful	neglect	worried
claim (verb)	failure	shirk	wrong

To be sure, we all have occasion to try to get the *idea* of a negative word across. In the last letter quoted, for example, the idea of *wrong* is clearly indicated in the sentence, "Unfortunately, our Maintenance Department will need a more permanent structure to handle the coils." In effect here the writer is telling the reader that the drawings of the latter company are not *right*. And yet the way in which the idea is presented is not negative: surely "more permanent structure" is a *positive* idea.

In your own writing you should try to develop the technique of using as many positive words as you can when you must state a negative thought. But you must be very careful not to make the negative thought positive. Once you can see that the thought is exactly what you want to communicate, then you can face the challenge of making your letters sound friendly and interested—positive.

Letter Language

The discussion of tone should make you more aware of individual words than you have ever been before. As you think about language, you should also become aware of a "trap" you can let yourself get into. Many years ago there developed a language that became widely adopted as correspondence language by a substantial majority of letter writers in the business world. People read letter after letter containing this language, and, believing it to be the language for correspondence, they began to use it themselves.

The "vicious circle" of that letter language is still with us today. To be sure that you are not a victim, look closely at the stereotyped words and phrases in the following list. Then look at the suggestions for making the thought sound more human, friendly, and positive. Finally, before turning from this chapter, compare the stereotyped language with that of the good specimens throughout these pages. What you see should convince you that no letter would be as effective if it contained any of the trite, negative, and often vague and even meaningless language in the list below.

Stereotypes to Avoid in Letter Writing

above (as in, "We refer you to your letter on the above subject"). Say "Thank you for writing us about our underpayment of your Invoice No. 3298." Always repeat the idea in preference to being vague.

According to our records   Be specific. Say, for example, "Invoice No. 989 shows" or "The guarantee expired on August 26."

Acknowledge receipt of   Be specific. Say, for example, "Thank you for sending us a sample of your new additive."

agreeable to   (as in "We would like to know if this is agreeable to you"). Be direct and concise. Say, "We would like to know if you agree."

along this line, along these lines   Be specific. Say "on this topic" or "as far as our testing program is concerned."

as per   (as in "as per our records"). See *According to our records.*

at the present writing   Prefer being direct and concise. Say, for example, "now," "right now," or, if appropriate, "today."

Attached please find   Write as if you were speaking. Say "Attached is," "We are attaching," or, if empathy is in order, "We are pleased to attach."

Be so good as to, be so kind as to   This is the way the Germans would say it, but "please" says it better.

contents duly noted   All this phrase means is, "Your letter was read immediately and carefully." It tells the reader nothing that he does not expect. He assumes that you read all your mail "immediately and carefully." Omit the phrase.

Enclosed please find   See *Attached please find.*

feel, felt   (as in "We feel," "It is felt") "Feel" is a verb indicating either touch or emotion. If you wish to tell the reader what your analysis is, say "We believe (conclude, think)" or "It is believed (concluded, thought)."

hand you   (as in "We hand you our proposal"). This act is an impossibility if the "handing" is done by mail. Say, instead, "Enclosed is our proposal," or, "We are glad to send you our proposal."

Hoping   (as in "Hoping that our report meets with your approval, we remain"). The participial phrase is weak except when it summarizes or subordinates an idea effectively. Such a phrase is always weak at the end of a letter. If you must say "I hope" or "We hope" there, be specific in completing the sentence. Say, for example, "I (we) hope that you approve our report."

immediate attention   You can use this phrase effectively when urgency is required. However, you should prefer being more specific—as you would be in writing, "We will ship your order immediately" and "We are shipping your order today."

in hand   (as in "Your letter is in hand"). All this says is "I am looking at your letter," and it would look pretty childish if it actually appeared in a letter. Say "Thank you for your letter" or something else that gets you into the message.

kind   (as in "Thank you for your kind letter"). The use of "kind" in this way is not very logical. Use a logical adjective—if at all—such as "friendly" or "welcome," or say, simply, "I appreciate your letter."

not in a position to   (as in "The company is not in a position to grant your request"). This is a way to avoid saying, "We cannot grant your request." If you mean, "Other employees would be upset if we allowed you more time off," you should say just that.

pending receipt of (as in "We shall reserve our decision pending receipt of your final report.") For variety, use something like, "We shall reserve our decision until we receive your final report." You should always use forceful, direct, and simple statements.

Permit me to   Say "May I" or "I'd like to." "Permit me to" is questionable today and to many it may seem patronizing.

Please refer to (as in "Please refer to my letter of October 12"). You should never use this phrase to make a reference when you begin a letter. The phrase gives the reader a command and therefore may seem to be cold and unfriendly. Say, for example, "I am writing again to remind you of the program we have arranged for you and your staff on October 12."

pursuant to (as in "pursuant to our earlier statement"). Say simply, "As we noted before," or make a similar direct reference.

re Prefer not using Latin in your letters. Say "about" or "concerning."

referring to (as in "Referring to your letter of December 6, we are glad to send you our prices for No. 76 generators in carload lots"). This phrase is acceptable if the clause following has a subject that "Referring to" modifies. If it does not modify the subject, "referring" is a dangling participle (as in "Referring to your letter of July 9, you will be pleased to learn, . . ."). Often used at the beginning of a letter, this construction is also questionable because it may seem mechanical and rather stilted. Always prefer making a direct statement: "Thank you for your letter of July 9. You will be pleased to learn. . . ."

same (as in "We must receive your shipment by the 12th. Ship same today"). Use the logical pronoun ("it" in this case) or noun that you would use in conversation.

Thanking (as in "Thanking you in advance for your kind attention to this matter, we remain"). Like *Hoping,* "Thanking" is a poor ending. The reader may also think that it is presumptuous because in most instances, saying "Thanking you" or "Thank you (in advance)" before your request has been granted may seem to be an imposition on the reader.

the writer   Say "I" ("we" if you are writing as the spokesman for your company). Your letters will be more natural if you refer to yourself (your company) as you do in speaking to others.

Trusting   See *Hoping* and *Thanking.*

We have your letter of   This is another meaningless beginning for a letter. See *in hand.*

we remain "We remain" is another expression that is out of date. As you consciously avoid using the participial endings (*Hoping, Thanking, Trusting*), make it a point not to use this archaism too.

without further delay   This phrase is not bad English, but, like "immediate attention," it is overused. Try to be more specific and emphatic, as you would be in saying "this week" or "within the next 10 days."

would state (as in "In reply, would state"). Lazy writers tend to write "telegram" English. As you know, however, a sentence must have both subject ("I") and verb. Furthermore, the verb "state" may seem very negative or pompous to the reader. Use "say" or some other verb like it.

Yours (as in "Yours of August 31 in hand"). "Yours" in this use means "your letter" and one may readily look upon such abbreviated usage as questionable.

Remember: avoid using words and phrases that appear in italics in the preceding list. Prefer using the language indicated in the commentary on each stereotype. Or, even better, express yourself in your own way.

EXERCISES

LAYOUT

1. Copy one or more of the letters on pages 158, 159, 162, and 163, but for each copy use a different layout from the one illustrated.
2. Eliminate any errors or violations of convention that you find in the following excerpts from letters.
   a. (Typed heading)

   ```
                              March 28, 19__
                              712 Harkness St.
                              Fullerton Ind.
                              Zip Code 46833
   ```

   b. (Inside address with attention line)

   ```
   Wilson Electric Company
   Millerstown, Wiscon., 53641
   Attention of Mr. R. G. Wheeler, Sr.
   312 Wilson Build.

   Dear Mr. G. R. Wheeler;
   ```

   c. (Heading for second page)

   ```
   2, Wheeler, January 8, 19__
   ```

   d. (Signature)

   ```
                              Sincerely Your's,
                              T. M. Frey
                              ELECTRONICS, INCORPORATED
                              Manager; Research Department
   ```

e. (Enclosure and attachment lines)

```
                              en. copy of R. Browns letter
                              at. Revised specifications
```

REQUESTS AND INQUIRIES, AND REPLIES

3. Discuss the effectiveness or ineffectiveness of the following letter of inquiry, which was written by a student for a technical writing course.

```
                              728 Washburn Ave.
                              Phila., Penn. 19088
                              Nov. 18, 19__

L.L. Warren
Dir., Res. & Dev.
U.S. Minerals Comp.
N.Y., N.Y., 10015

Gentlemen:

    A student at Eastern University, I am doing research on the
different conditions and concentrations of sulfur dioxide fumes
in stack gases.  Anything you could send me in the form of pub-
lished reports or brochures on this subject I would consider in-
valuable information towards the completion of my report.  I am
particularly interested in equipment available for extraction.
On the extraction of sulfur dioxide from stack gases I would
particularly like to know: what percentage of sulfur dioxide is
removed by the various equipment you produce, the cost per pound
of sulfur dioxide processed to operate this equipment, and is
adsorbtion or absorbtion employed in the sulfur dioxide removal.
Because my report is due November 24, I must have your answer by
November 21.


                              Sincerely Yours,

                              Robert J. Bogard
```

4. Discuss the effectiveness or ineffectiveness of the following reply to a letter of inquiry. The reply is written in answer to the letter quoted in Exercise 3.

```
Dear Mr. Bogard:

    I regret that I am unable to send specific answers to the
questions raised in your letter of November 18.
    I am sending some reports and brochures, however, as you
requested.  Several of these are specifically concerned with the
subject of your report; others describe the many other kinds of
equipment we manufacture.
```

I am answering immediately in order to help you meet your deadline. Good luck with your report.

Cordially,

L. L. Warren, Director
Research & Development

ORDERS AND INVITATIONS TO BID

5. Discuss the effectiveness or ineffectiveness of the following invitation to bid, the writer's company's first contact with the supplier.

Gentlemen:

Please send your quotation on supplying us, all freight charges prepaid, the following items for delivery no later than April 17:
Six (6) gross lead pencils carrying our imprint.
Twelve (12) dozen tablets for office use.
Twelve (12) gross pens bearing our special imprint, to be distributed to customers and prospective customers.
Six (6) boxes paper clips for office use.

Very truly yours,

CLAIMS AND COMPLAINTS, AND REPLIES

6. Discuss the effectiveness or ineffectiveness of the following claim letter.

Gentlemen:

We are in receipt of your invoice No. 81835, dated June 20. However, we cannot understand why you are charging us $85.00 per length for four 25-foot lengths of Teflon Fluid Hose No. 204-79. The latest price that we have shows this price to be $85.00 per 100 feet.
We therefore expect to receive by return mail your credit memorandum to correct your error.

Very truly yours,

7. Discuss the effectiveness or ineffectiveness of the following complaint letter.

Gentlemen:

We are completely dissatisfied with the explanation in your letter of August 21 regarding your invoice No. 1936 dated July 15 in the amount of $1,176.00.

You charged us $1,132.00 for 1,000 feet of 500 MCP NRWP wire that should have cost us only between $850.00 and $900.00. You claim that you paid a premium for this material in order to complete a rush delivery to our plant within 24 hours. However, we feel that a premium of approximately 30% is entirely too much.

While we hesitate to say that we will penalize your company if this situation should occur again, we must protect our purchasing position.

Next time, phone us before you pay a premium to meet our delivery requirements.

Sincerely yours,

NOTICES, ANNOUNCEMENTS, AND OTHER INFORMATION-GIVING LETTERS

8. Discuss the effectiveness or ineffectiveness of the following notice.

Gentlemen:

We have received your acknowledgment, dated May 14, of our order TN-50761 covering four tankcars of liquid caustic soda.

Your acknowledgment indicates that you intend to ship one car each on June 7, 13, 16, and 23.

Please note that our order specifies that each car be here in our plant on the date listed.

We will appreciate your revising your shipping schedule and advising us the precise date you intend to ship.

Very truly yours,

TONE AND LETTER LANGUAGE

9. Discuss the effectiveness or ineffectiveness of the tone and language in the following excerpts from letters; then rewrite each excerpt.
   a. We cannot explain the malfunction because you failed to return the valves for our inspection.
   b. To complete the questionnaire you must place a check in the box beside the phrase that most accurately represents your answer.
   c. Please be advised that, according to our records, your check was forwarded to you two weeks ago.
   d. We feel that we can accommodate your request as soon as we are in a position to resume production.
   e. Referring to your letter of April 29, you should know that, pursuant to our agreement, your order has been given our immediate attention.
10. Discuss the effectiveness or ineffectiveness of the tone and language in the following complete letters.

a. Gentlemen:

    We have in hand your letter of June 13 offering to accept the return of the 6-inch gate valve that we ordered in error under our purchase order No. 71620.
    Please be advised that we have decided to keep this valve for use on a future job.

                                            Very truly yours,

b. Gentlemen:

    Please be informed that your attached acknowledgment of the subscription covered by our order No. 8051 is incorrect insofar as the mailing address is concerned. Your records should be corrected to ensure that the publication is mailed to the following address, as specified in our original order:

                    Mr. George W. Simpson, Supervisor
                    Wilton Electronics, Incorporated
                    9173 Industrial Park
                    Warren, North Carolina 13091

                                  Sincerely yours,

SPECIAL PROBLEMS

11. Write a simple request to a company or other organization, to obtain more information about a product or service, etc.
12. Write a multiple-question inquiry to a specialist who has done a great deal of work on the subject of a research project you are beginning.
13. Rewrite the letter quoted in Exercise 3. Make any changes that you think are necessary to improve the organization, paragraphing, tone, etc.
14. Rewrite the letter quoted in Exercise 5; make any changes that you think are necessary to improve this invitation to bid.
15. Rewrite the letter quoted in Exercise 6. Make any changes that you think are necessary to make the claim more effective.
16. Rewrite the letter quoted in Exercise 7. Make any changes that you think are necessary to make the complaint more effective.
17. Write a claim or complaint letter about an item that you purchased recently and that you think is less acceptable than advertising for the item indicated.
18. Rewrite the letter quoted in Exercise 8. Make any changes that you think are necessary to make the notice more effective.
19. Rewrite the letters quoted in Exercise 10. Make any changes that you think are necessary to make the letters more effective.
20. Make a study of at least ten business letters you and your friends have recently received. Then write a report evaluating the paragraphing, organiza-

tion, tone, etc., of these letters and drawing any overall conclusion(s) that are indicated by your evaluation. Attach the letters to your report as exhibits, and also quote from them to make your points clear and convincing.

Chapter Nine

MEMOS AND
INFORMAL REPORTS

This chapter covers typical messages that are written to keep an organization operating as a smooth, efficient unit. These messages—memos and informal reports—are called informal to distinguish them from formal reports and proposals, which are analyzed in Chapter 10. Informal messages differ from formal reports in physical presentation.

Both the memo and the informal report can be defined quite literally. The memo is an intracompany letter which is used by workers to objectively present requests, inquiries, etc., to other workers. Literally, the memo is a message "to be remembered." The informal report is an intracompany accounting for regular or special work that has been done. Literally, the report is a "bringing back" of facts, conclusions, and, where appropriate, recommendations.

As the examples on the following pages illustrate, memos and informal reports are presented in a form that can easily be distinguished from the letter. And both, because of the need for objectivity, are much more matter-of-fact and less empathic than extracompany correspondence. The one exception is the report sent to a client in letter form. In the letter report, empathy must be used—as indicated and illustrated in Chapter 8.

Memos and reports are sent in all "directions" within an organization. They are written by workers for leaders, by workers for workers on the same level, and by leaders for workers.

MEMOS

Informal Proposals

Intracompany proposals range from the handwritten note that goes into the office suggestion box to more involved and complex messages. Since the note for the suggestion box is impromptu, we shall look closely at more detailed proposals.

Proposal to Make a Change

Efficient workers continually try to find ways to improve the operation of their unit or department. Aware that any improvements can benefit themselves—indirectly if not directly—they submit proposals for a change in methods, procedures, equipment, etc. The following proposal, for a change in specification, is typical.

Date: June 30, 19___

From: T. E. Smith

To: Mr. F. B. Tracy, Manager, Engineering

Subject: Change in Specification for Pump Impeller Metal

We request your approval of our rewriting the specification for the metal used to construct centrifugal pump impellers that are subject to corrosion in their normal operation.

At present, the specification in Volume VII, Section 17 of our Design Specifications manual reads:

All metals shall conform to ASTM A-186, Revision No. 3.

In addition, these metals shall be prepared or machined according to the method described under Procedure No. 4 in the ASME Manual.

Our inspections during the past year have shown that ASTM A-186 is a poor alloy for impellers under corrosive conditions. A better specification was outlined in a new method for preparing the metal, which was introduced last month at the North American Institute for Engineering Practice held at White Sulfur Springs. A copy of this method is attached.

This method of preparing the metal is superior in every way to that in Procedure No. 4. Especially significant is the point that the strength of the metal is not reduced during treatment. As you know, under the method described in Procedure No. 4 the strength is reduced.

Please let us know if we may proceed with rewriting the specification.

Note the organization of this request. Since the writer does not consider the subject controversial, he begins immediately with the request. Then he presents information about the present specification and goes on to justify rewriting the specification.

Naturally, some subjects for proposals *are* controversial. That is, they are questionable changes, require great expenditures, etc. Following is a proposal for change that some leaders might have misgivings about, if two major points were not made.

```
Date:     April 17, 19__

From:     P. C. Murray

To:       Mr. J. S. Thurston, Superintendent of Operations

Subject:  Analyzing the Metals Content of Catalytic Charge Stocks

          For years we have been using the colorimetric
          method to analyze the metals content of catalytic charge
          stocks.  This is a very accurate method for determining
          the amount of vanadium and nickel.  However, as you may
          recall, it has two major disadvantages, one of which
          accounts for the other.  It requires 5 hours' working
          time and 28 hours overall; and because of the excessive
          time involved, we accumulate samples and test them only
          once a month.

          Recently, we worked out a technique using the 100 KV
          X-ray spectrograph.  With this instrument we can ade-
          quately determine the amount of vanadium.  The spectro-
          graph is not sensitive enough to determine the nickel
          content; but because the vanadium-to-nickel ratio is
          constant generally, we can eliminate the nickel deter-
          mination for most routine control samples.

          The advantages of the spectrographic method are ob-
          vious.  We can complete an analysis for vanadium in 5
          minutes, and we can test samples frequently--daily if
          necessary.  For these reasons, we think it will be
          highly advantageous to use the spectrographic method
          instead of the colorimetric.

          If you approve, we plan to test the catalytic charge
          stock streams from No. 86 and No. 87 Stills on Monday
          and Wednesday of each week.  If you wish, for a month or
          two we can also use the colorimetric method as a check
          on the accuracy of our results.
```

The important points made in the above memo are as follow:

1. The metals content of the catalytic charge stocks can be determined much more often if the spectrographic method is used.

2. For a month or so, the colorimetric method can *also* be used—as a check on the accuracy of the results obtained with the use of the spectrographic method.

When the subject of your own writing is controversial, you will do well to develop the proposal in the same way as the one we just presented.

1. Review the problem (present method, procedure, design, etc., and point out its shortcomings).
2. Present and discuss the method, etc., that experimentation and analysis have shown to be more efficient, economical, etc.
3. Clearly show the advantage of what is being proposed.
4. Indicate how what is being proposed can be implemented so that all who are concerned will be satisfied.

To be included in this outline is the cost of making a change when extra expense is involved. In the example just quoted, cost was not a factor because the equipment needed was already available. However, you frequently may have to supply details on the amount of money that will be needed to make the change requested. Naturally, a justification for the expenditure should be presented at all times. Remember that leaders need to be convinced before they will approve a proposal for change.

Proposal to Carry Out Research

Not as common but very important to the successful operation of an organization is the proposal to carry out research. Unlike the proposal to make a change, the proposal to carry out research is always based on some "unknowns." For example, the time and the cost of future experimentation can only be estimated. However, like any hypothesis, a proposal for research is based on either prior knowledge and experience or a comprehensive accumulation of data. In most cases, the estimates are, therefore, quite accurate.

Note, for instance, the amount of information that the writer of the following proposal[1] had gathered beforehand. His subject is the development of a new product, galvanic zinc paint, from raw material already being manufactured by his company.

This illustration helps us to see how important it is to have reliable data on which to base estimates. Complete coverage of the subject, and, of course, precision are always necessary.

A comprehensive outline of the proposal for a research project of this kind is presented and discussed in Chapter 10. Special emphasis is given there to preparing a proposal for customers and potential customers.

---

[1] In some companies, a presentation such as the example on the following pages would be called a formal proposal. "Add a cover, title page, transmittal, and table of contents," these companies would say, "and the presentation would indeed be a formal proposal." Thus your instructor may frequently refer to the Consolidated Zinc proposal when he covers the analysis of the formal proposal on pp. 235–41.

CONSOLIDATED ZINC COMPANY

Intracompany Correspondence

Date: January 5, 19__

From: L. A. French, Paint Laboratory

To: Dr. John B. Nathan, Director of Research

Subject: <u>Galvanic Zinc Paint--Proposal for a Research Project</u>

INTRODUCTION

<u>Purpose and Scope</u>

The purpose of this project will be to develop information needed to begin the manufacture of galvanic zinc paint.

The scope of the project covers two phases:

Phase I, Market Development, will cover gathering data on the cost of building and equipping a plant, the cost of manufacturing paint, and the major markets (with details including the types and amounts of zinc-dust paint consumers use, and the price they pay for paint).

Phase II, Research, will cover the cost of refining our existing formulas or of reformulating to meet customers' specifications. Phase II will not begin until Phase I is completed.

This is a proposal for an initial exploration, based on raw-material, production, and equipment costs previously estimated by the Paint Laboratory.

<u>Background</u>

The use of highly pigmented galvanic zinc paints has been growing rapidly in the United States during the last ten years. Automobile manufacturers are probably the largest single market at present. Other volume uses include primers for steel construction and primers for ships. A 45,000-ton tanker requires 100 tons of zinc-rich paint for priming; this amount of paint contains approximately 85 tons of zinc dust. Galvanic zinc paints also have been suggested for coating concrete reinforcing rods, a use that would mean a substantial increase in the size of the market.

Statistics clearly show that the market is increas-

ing. Seven years ago, an estimated 2,000 to 3,000 tons
of zinc dust were used for zinc-rich paints. This figure
rose to an estimated 4,000 to 6,000 tons five years ago,
and to 10,000 to 14,000 tons two years ago.

For several years, we have been making zinc dust for
in-plant consumption. The production process was developed with a view to eventually marketing a zinc-dust pigment; and production personnel are now confident that
they can control particle size and zinc oxide content
enough to manufacture an acceptable pigment. As a result, management has asked for raw-material and manufacturing costs for galvanic zinc paints made with the zinc
dust in its refined state.

The Paint Laboratory has formulated and tested
organic-vehicle galvanic zinc paints, and compared these
with existing commercial and inorganic zinc-dust paints.
A sample of the paint made for the automotive industry
appears to be based on an epoxy ester vehicle. We have
been able to match the physical properties and performance of this paint quite closely. We have also used
successfully a chlorinated rubber/chlorinated paraffin
vehicle.

PHASE I, MARKET DEVELOPMENT

Research

There will be no research for Phase I.

Market Development

People

The professional services of the Zinc Oxide Technical Service Department will be needed for 12 man-days.

Services

A market survey will be needed to determine potential customers, their volume requirements, and the type
of galvanic paint they use. For much of this information, the services of a consultant familiar with the
uses and distribution of zinc-dust paints will be needed.

Time

Approximately two months will be required for completing Phase I.

2

Costs

Consultant's Fee (G. K. Hughes)	$ 750.00
Zinc Oxide Technical Service Expenses	750.00
Total	$1,500.00

PHASE II, RESEARCH

General

Research for Phase II will not begin until Phase I has been completed.

The data obtained from Phase I will determine the amount of research needed to produce one or more galvanic zinc paints suitable for industrial use. All potential customers will require data on the corrosion resistance of the product, especially its resistance to salt spray and humidity. Many consumers will want to know the particle size of the dust used. The ability of galvanic zinc primers to weld is required by the automotive industry. This is a costly test, performed with specialized welding equipment, so we must assume that the customer will conduct his own welding tests. If our formulas are similar to existing paints in the physical properties that influence the ability to weld (pigment volume, zinc content, zinc-dust particle size, type of binder, etc.), we assume that their performance will be similar.

Special Requirements

People

	Man-Hours
Professional	
Paint Laboratory	200
Microscopy Laboratory	50
Support - Technician, Paint Laboratory	500
Total	750

Facilities

Salt Spray Cabinet
Humidity Chamber
Spray Booth

Services

No outside services will be needed for Phase II.

3

Time

Eight months will be required for the second phase. About six months will be needed to correlate salt-spray and humidity-chamber tests with corrosion tests previously made in the Paint Laboratory, and to match the performance of currently acceptable paints.

Costs

Labor		$4,000.00
Equipment		3,850.00
	Total	$7,850.00

ECONOMICS

The estimated market volume and selling price will be determined by the market survey, Phase I. Our preliminary cost studies, however, are based on two levels of production: (1) incorporating 1,000 tons of zinc dust per year into 90,000 gallons of paint, and (2) incorporating 2,000 tons of zinc dust per year into 180,000 gallons of paint. A selling price of eight dollars per gallon to automobile manufacturers--which was obtained from our Detroit agent--was used as the basis for calculating profit.

Manufacturing Cost

	90,000 gal./yr.	180,000 gal./yr.
Equipment	$42,785.00	$54,880.00
Building	87,770.00	87,770.00

Labor

90,000 gallons/year	$24,400.00
180,000 gallons/year	48,000.00

Overhead

(Heat, light, power, telephone, office supplies, office work, fire insurance, taxes, maintenance)

90,000 gallons/year	$ 9,600.00
180,000 gallons/year	12,000.00

Material Cost

(Including container and shipping charges)

90,000 gallons/year	$472,000.00
180,000 gallons/year	945,000.00

4

<u>Sales Cost</u>

The sales cost is estimated at $0.20 per gallon.

<u>Unusual Technical Service Costs</u>

It is estimated that approximately $6,000 will be needed for technical service costs during the first twelve months of production.

———

<u>Profit</u>

<u>On 90,000 gallons/year</u>	$189,000.00
Add incremental profit (1,000 tons SHG zinc dust)	87,000.00
	$276,000.00
Subtract incremental profit (1,000 tons metal)	125,000.00
<u>Net gain before taxes</u>	$151,000.00
<u>On 180,000 gallons/year</u>	$392,000.00
Add incremental profit (2,000 tons SHG zinc dust)	174,000.00
	$566,000.00
Subtract incremental profit (2,000 tons metal)	250,000.00
<u>Net gain before taxes</u>	$316,000.00

The incremental profit for HG dust is $20 per ton less than for SHG dust. However, this difference would be canceled by the lower raw-material cost of paint containing HG dust.

<u>Years to Pay Out</u>

To pay out, two and one-half years are needed for 90,000 gallons/year of paint; one and eight-tenths years for 180,000 gallons/year of paint.

<u>Profitability Index</u>

For 90,000 gallons/year	31%
For 180,000 gallons/year	44%

5

These assumptions were made in the analysis of economics:

1. In the determination of the profitability index:

   a. The consultant's fee and technical service costs for a market survey ($1,500) were charged to the costs for the first operating year.
   b. The labor for research to develop a competitive paint ($7,850) was charged to the costs for the first operating year.
   c. The equipment for paint research was charged off as part of the equipment cost for the paint plant.
   d. Unusual technical service costs ($6,000) were included in the overhead for the first two years of production.
   e. Profit figures were based on producing paint for four months this year at 50% of the ultimate production rate; producing paint all next year at 65% of the ultimate production rate; and producing paint thereafter at 100% of the ultimate production rate.

2. The full production of 90,000 gallons or 180,000 gallons of paint per year can be sold after 16 months of operation.

3. The raw-material costs of products that appear satisfactory in our laboratory will closely approximate the costs of products acceptable to the trade.

4. The selling price of galvanic zinc paint will reflect changes in the cost of zinc dust.

5. All estimates were based on the use of our own dust, and costs were based on the market price of zinc dust prevailing at the time of computation.

ESTIMATES OF SUCCESS

Estimate of technical success: 90%

Estimate of economic success: 80%

6

Not all proposals for research are as complete or precise as the one on the preceding pages. At times, for example, a proposal is submitted to solve a problem that has not been fully defined. Following is an illustration of this type of proposal.

Date: August 18, 19__

From: Lawrence F. Billings

To: Mr. C. V. Allen, Manager of Operations

Subject: <u>Cement-Dust Control</u>

     The county's new dust-control ordinance specifies a limit of .5 lb of dust per 1,000 lb of standard stack gas. For months we have tried to improve our control of the amount of dust escaping from our stack. However, we have not been able to do so with our limited facilities and time. Our dust figures are still more than twice as high as the limit.

     The deadline for complying with the ordinance is January 1. It is obvious to us now that we will need extensive research on the subject in order to meet the deadline. Such a study would include a search for answers to these major questions:

1. What type of dust-control equipment will best suit our needs?

2. How effective is the best dust-control system?

3. Is there enough space available in the plant for installing the best system?

4. What will be the initial capital investment for the equipment recommended?

5. How long will it take to install the system selected?

6. What will be the annual operating cost of the system?

7. Can we recover and make use of any material collected by the system?

     To obtain the answers to these questions, we would cover the literature available on the subject in recent journal articles, examine reports from companies like ours that have already installed dust-control equipment, and compare costs and other details provided by the manufacturers.

     Because of the deadline, we believe that favorable action on this proposal should be taken at once. Since members of our staff are willing to put off their vacations until late fall if necessary, the research can be carried out full time without an adverse effect on the department's regular work.

As this proposal shows, important details such as estimated costs need not always be included. When emergencies arise and problems must be solved right away, estimates can be deferred until later. What matters most in these cases is that the project be approved as soon as possible.

As a student, you probably will write proposals having some "unknowns." A complete analysis of a proposal such as the one just quoted shows that you will do well to follow this outline:

1. Introduce the problem, presenting the important facts and conclusions.
2. Indicate generally what needs to be done; then either present specific questions to be answered or list the individual factors of the problem. (Such an outline is called the *scope,* or areas to be covered, of the problem. Specific and detailed, the scope shows exactly what the writer thinks makes up the problem. Of course, it also indicates what will *not* be covered. In other words, the outline of scope limits (restricts) the problem. If later it is found that questions (factors) should be changed—added to, subtracted from, or modified from those first presented—then the scope will change (and perhaps the problem, too).
3. Indicate what sources of information will be used to answer the questions (account for the factors) that make up the total problem.
4. Include any qualifying statements that leaders will need or want to see, and motivate leaders to approve the project proposed.

The sample formal report illustrated in Chapter 10 shows that the proposal just presented was approved, and a solution to the problem found. (See pages 221-35.)

Transmittals

Often when a study has been completed, the writer will include what is called a transmittal, or covering memo, with his report.[2] This memo introduces the report to the reader. Sometimes the transmittal says little more than, in effect, "Here's the information you asked for." For a study like that just referred to, however, a longer transmittal is usually needed and expected. As the example (see page 224) illustrates, frequently the longer transmittal includes the writer's professional comments and some interpretation of what is being transmitted.

Procedures

Vital to the effective functioning of a laboratory is recording procedures for future reference and use. Such a record is illustrated below.

Note in this memo that the language and abbreviations used are appropriate. The procedure described is obviously intended only for the participant—the one who will actually do the experiment. The procedure is not intended for an observer.

---

[2] Called a "covering letter" in a report to a client, of course.

Date: November 15, 19__

From: Paul C. Murphy

To: Dr. Francis L. Boyd, Director of Research

Subject: Tentative Procedure for the Spectrophotometric Determination of Nickel

Introduction

This procedure can be used to determine nickel in zinc and cadmium metal.

The sample should contain between 10-100 micrograms of nickel.

The level of impurities in zinc and cadmium metal is generally too low to interfere with the method. Concentrations of diverse elements which do not interfere are as follow: 1 gram of cadmium; 100 milligrams of cobalt; 25 milligrams of copper; 1 gram of iron(III); 25 milligrams of manganese; and 1 gram of zinc. (Note 1.)

Summary

The sample is digested with a mixture of hydrochloric and nitric acids. Sulfuric acid is added and the sample is evaporated to strong fumes of SO_3. Interfering elements are complexed with citrate. The pH is adjusted to 7.5 to 10.0. Dimethylglyoxime is added, and the resulting nickel complex is extracted with chloroform. The organic phase is then washed twice with a dilute ammonia solution. The absorbance of the nickel dimethylglyoximate complex in chloroform is read at 375 millimicrons.

Apparatus

A Beckman Model B spectrophotometer with 1 cm borosilicate cells was used for this procedure.

Reagents

Standard Nickel Solution (10 g/ml) Dissolve 1.0 g of pure nickel metal in nitric acid and sulfuric acid. Take to fumes of SO_3. Cool and dilute to 100 ml. Pipet 1 ml of this solution into a 1-liter flask and dilute to volume with distilled water.

Dimethylglyoxime Solution (0.1%) Dissolve 0.10 g of the sodium salt in water and dilute to 100 ml.

Ammonium Citrate Solution (40%) Dissolve 400 g of ammonium citrate in water and dilute to 1,000 ml.

Dilute Ammonium (1:50)   Dilute 20 ml of concentrated ammonium hydroxide to 1,000 ml.

Procedure

1. Weigh a sample of such size that it contains between 10-100 mg of nickel.

2. Dissolve the sample with 5 ml HCl and 5 ml HNO_3 acids. Add 5 ml of sulfuric acid and take to strong fumes of SO_3.

3. Cool, add 20 ml of water, and warm if necessary to get the salts into solution.

4. Add 25 ml of ammonium citrate solution.

5. Add several drops of phenolphthalein indicator. Neutralize with concentrated ammonium hydroxide until the solution turns a light pink (pH 7-10) and add 3 drops in excess. (Note 2.)

6. Transfer the solution to a 125-ml separatory funnel, using a minimum quantity of wash water, and add 5 ml of dimethylglyoxime solution.

7. Adjust the total volume to approximately 75 ml.

8. Pipet in exactly 10 ml of chloroform and shake for 1 minute.

9. Wash the chloroform layer with two separate 10-ml portions of (1:50) ammonia solution. Shake vigorously for exactly 15 seconds each time.

10. Transfer the organic layer in a 15-cm centrifuge tube and centrifuge.

11. Transfer a portion of the organic layer to a 1-cm spectrophotometer cell and measure its absorbance at 375 mμ. (Note 3.)

Notes

1. If these limits for cobalt and copper are exceeded by a factor of 10, increase the quantity of glyoxime added by the same factor.

2. For highly colored samples, adjust the pH to 8.3 with NH_4OH and add 3 drops in excess.

3. Prepare the calibration curve by adding known amounts of nickel from 0 to 100 μg to 150-ml beakers. Carry through the above procedure, starting with step 1. Plot A vs. μg Ni.

Discussion

The presence of much manganese (25 mg) tends to inhibit
the extraction of nickel. Iron should be in the ferric
state, as the ferrous state forms a dark chloroform ex-
tract. Copper(II) and cobalt both form complexes with
glyoxime which are extracted to a small extent and im-
part a brownish coloration to the chloroform layer.
They are transferred completely to the aqueous phase
when the chloroform layer is washed with dilute ammonia.
Hence, enough dimethylglyoxime must be added to react
with diverse ions and to leave an excess for nickel.
The extraction of nickel is complete in the pH range of
7.2-12 for a citrate medium. Above pH 12, a brown oxi-
dized nickel dimethylglyoximate complex is formed.

Oxidizing agents must be absent because they will
oxidize the nickel complex. Oxidation results in the
incomplete extraction of nickel.

The absorption curve of the nickel complex in
chloroform shows a maximum at 375 mμ. Beer's Law is
obeyed, and the color is stable for at least one day.

Note also the differences in style. The third-person (declarative) mood is used for the introduction, summary, apparatus, and discussion section; the second-person (imperative) mood is used for the "how to" reagents, procedure, and notes sections. The differences are necessary, of course, to explain and to direct.

Finally, note the logic of organization. After introducing the subject of his procedure memo, the writer tells the participant-reader what, in brief, occurs during the procedure. Then, after the summary of procedure, he tells the reader what apparatus he needs, how to prepare the reagents to be used, how to pro-ceed, and what to take special note of. Then, under the discussion section, he supplies interpretations that will help his reader know what to look for (and look out for) as he follows the procedure.

INFORMAL REPORTS

Like formal reports, informal reports are written to fulfill a variety of "account-ing for" purposes. Essentially, however, reports have one of two functions: they present only facts for the most part,[3] or they present facts and interpretation. Reports that present "facts only" are called *informational*. Their chief function is to tell readers (who),[4] what, when, where, and how. They present statistical

---

[3] See the discussion of generalizations of fact, page 195.
[4] Naturally, in a great deal of technical writing, *who* does the work is of less concern to the reader than the work itself.

and other data, definitions, descriptions, and other kinds of explanation. Detailed accounts of a machine's components, a production process, another person's theory, the progress of a meeting, the stages of an experiment, etc.—presented objectively, of course—are all examples of informational reports.

Reports that present "facts and interpretation" are called *interpretive*. In addition to supplying the facts of (who), what, when, where, and how, interpretive reports tell readers why, how much and how little, how good and how bad, and what to do and what not to do. They present ideas, hypotheses, original theories, and—especially—conclusions, evaluations, and recommendations.

In brief, informational reports tell readers what they need or desire to know; interpretive reports, what they need or desire in order to make decisions and take actions.

Informal reports vary greatly in length and form. In almost all organizations, at least a few reports are one-page-or-less, fill-in forms. In general, however, reports cover more than one page and are presented in a form that each author determines is appropriate to his purpose. Characteristically, the two basic structures for informational and interpretive reports are as follow:

Informational Reports	*Interpretive Reports*
SUMMARY (if a long report)	SUMMARY (if a long report)
INTRODUCTION	INTRODUCTION
FACTS ONLY	FACTS AND INTERPRETATION
APPROPRIATE ENDING	CONCLUSIONS
APPENDIX	RECOMMENDATIONS
	APPENDIX

A summary of a page or less should always be included if the report is long or has many details. ("Long," as defined by many organizations, is four pages or more; "many details" means important specific data taking up a half page or more in reports of two pages or more.)

The summary and the introduction can be interchanged. If the summary comes first, then part of it must be a summary of the introduction. If the introduction comes first, no part of the summary should cover it; the assumption is that readers will read the introduction first.

Like all good writing, "facts only" reports should have topic sentences (summary sentences for paragraphs) throughout. These, of course, cover the idea of each paragraph in the report. Thus, they are called "generalizations of fact." Examples of topic sentences for "facts only" (informational) and "facts and interpretation" (interpretive) reports help you to see the difference in the roles they play:

Topic sentence for an informational report

```
     The laboratory uses a 15-minute heat test to estimate the
total solids and the detergency rating.
```

Topic sentence for an interpretive report

```
    The laboratory's 15-minute heat test is better than any other
test for estimating the total solids and the detergency rating.
```

In the first illustration, we find only a summary statement of fact. We expect the rest of the paragraph to supply a factual description of the test. In the second illustration, we find an evaluation. We expect the rest of the paragraph to supply factual support showing that the laboratory's test is better than any other test.

Informational reports often are ended most effectively with the last set of facts to be presented. Such an ending is especially appropriate when a summary precedes the report. At times, however, a short paragraph briefly summarizing the point of the report, emphasizing accepted applications of the subject, or explaining the history, proven popularity, or a special feature, may be meaningful to the reader. A good rule of thumb is to add something only if you can justify it.

If only for emphasis, conclusions and recommendations should be added to many interpretive reports (for example, putting conclusions and anticipated solutions in their proper perspective). Such emphasis is *always* desirable when a decision is to be made after the leader has read the report. The value of terminal conclusions and recommendations increases with the length of the report—even when the report has an introductory summary.

An appendix is best thought of as a "catch-all." The appendix is the place for endnotes (if they are used), the bibliography, charts, graphs, computations, and other material *that is supplementary in nature.* If anything is more than supplementary, it belongs in the body of the report, not in the appendix. A point to remember is that few readers look closely, if at all, at an appendix. The rule of thumb here, then, is to put in the appendix any material that cannot logically and effectively be put in the report. Appendix materials satisfy curiosity—not *needs.*

A Specimen Interpretive Report

To illustrate the points made on these pages, an interpretive report based on both field and laboratory research is presented. Like any other data-based investigative report of merit, this one clearly separates the factual results from their interpretation. A complete outline for a "laboratory" report that you may have to write is as follows:

Summary (including or limited to conclusions and recommendations—"Conclusion and Recommendation" in the specimen)

Introduction (clearly stating the purpose and scope of the investigation)

Review of Previous Investigations (presenting perhaps summaries of the work of other investigators of the same or a similar problem—covered in the Introduction in the specimen because it is brief)

Present Experimental Work (showing details of the apparatus selected and the procedures followed—not needed in the specimen except to show later, under the heading "STEPS TAKEN TO SAVE OIL," a last-hope attempt to put off having to make the oil change)

Results (presenting in detail the specific results obtained in the study—in the specimen, the results just derived are compared meaningfully, in table form, with those obtained for the same purpose at an earlier date)

Discussion of Results (interpreting the individual important facts uncovered by research—presented in the specimen in the paragraph following the table and supplemented by references to exhibits for the curious)

Conclusions (interpreting the discussion of results to explain why what happened, happened—covered in the specimen as an interpretation based on the results noted in a similar investigation on a sister ship)

Recommendations (presenting a solution to the problem investigated—shown in the specimen as a list of procedures to be followed in order)

Appendix (showing in tables, graphs, etc., what has already been reported and interpreted—consisting in the specimen of a comprehensive table of data and three charts as reinforcements for the facts, conclusions, and interpretations presented in the report proper; note that not all the data are needed for the conclusions and interpretations, but that all are presented to show that the investigation was thorough)

AMALGAMATED OIL COMPANY

Intracompany Correspondence

Date: August 7, 1971

From: C. F. Matthews

To: Mr. P. T. Shelley

Subject: Replacement of Turbine Oil in SS Hunter

CONCLUSION AND RECOMMENDATION

Analysis of both samples of the oil obtained from the main turbine lubricating system of the SS Hunter shows that the oil is deteriorating rapidly and should be replaced at once.

Specific recommendations for making the change are presented on pages 2-3.

INTRODUCTION

In the routine check, made on May 23, we found that the oil in the main turbine lubricating system of the SS Hunter had deteriorated considerably since March 25, when the last check was made. Hence, four days ago,

when the Hunter returned from its last voyage, we obtained another sample for laboratory testing. This report presents the results of the last two analyses; the
May 23 sample of oil is labeled R-3,847; the August 3
sample, R-5,380. (Attachment A shows all test results.)

COMPARISON OF IMPORTANT LABORATORY RESULTS

Test	Sample R-3,847 May 23	Sample R-5,380 August 3
Color, ASTM D 1500-58T	7.5	8.0
Appearance After Agin 24 Hr at 70-80°F	hazy	bright
Water: PPM	320	42
Neutralization Value ASTM D 974-58T Total Acid No.	0.27	0.47
Agent 046: %	0.52	0.35
Ash (Max Temp., 1,000°F): % Analysis Spectrographic: PPM Copper	10	15

Inspection of sample R-5,380 showed that the total
acid number increased 43% over the previous sample,
R-3,847. (See Attachment B.) The soluble copper content increased 33%. The haziness is attributed to 320
PPM water in the oil. The color increased 0.5 unit in
approximately two months. (See Attachment C.) The amount
of Agent 046 decreased from 33%. (See Attachment D.)

CAUSES OF OIL DETERIORATION

Previous experience with turbine lubricating systems
(most recently the system in the SS Gibson) shows that
the causes of the rapid deterioration of the oil are as
follow:

1. Rust preventives and contaminants were in the
lube oil system when the initial charge of oil
was made. Contaminants caused the oil to have
poor emulsion characteristics, short oxidation
life, and a dark color.

2. Copper had a catalytic effect on the oil in the
lubricating system. (Note: the oil lines to
and from the governor control mechanism on the
SS Hunter are 3/4 inch high-pressure copper
tubing.)

3. The present location of the suction line to the lube oil purifier from both lube oil gravity tanks is seven inches from the bottom. This condition allows approximately 50 gallons of water and extraneous matter to be trapped in the bottom of each tank.

STEPS TAKEN TO SAVE OIL

Addition of 0.5% by weight of Agent 046 was made to both samples of used turbine oils from the main lubricating system. However, the addition did not increase the oxidation life enough to warrant the use of a pilot solution.

SPECIFIC RECOMMENDATIONS FOR CHANGING THE OIL

The oil required for the change will be 8,000 gallons (6,000 for the system and 2,000 for the reserve oil storage). The steps to follow, in order, are these:

1. Remove used oil from the system by pumping it into a bunker tank. (This will be approximately 6,000 gallons.)

2. Drain all traces of oil from the sump tank, lube oil coolers, lube oil pumps, piping, and gravity tanks.

3. Inspect lube oil gravity tanks and the sump tank for any extraneous matter and clean by hand wiping.

4. Use the reserve oil in the forward storage tank (approximately 1,000 gallons) as a displacement oil for flushing and cleaning the system. The flushing operation should take approximately 4–8 hours at 130–140°F.

5. Repeat steps 1, 2, and 3.

6. Obtain samples of the new oil before charging and also after the oil has been circulated in the main lubricating system.

These procedures were followed in making the turbine oil change on the Hunter's sister ship, the SS Gibson.

If these recommendations are acceptable, the change can be made when the SS Hunter docks in two weeks. Laboratory personnel will be glad to assist the Marine Department in carrying out the recommendations.

Attachments 4

(A)   <u>PROPELLING UNIT TURBINE OIL ANALYSIS</u> August 7, 1971

Vessel:	<u>SS Hunter</u>	
Voyage:	48	50
Unit Make and Type:	Richardson Westgarth Double Reduction Geared Turbine, 13,750 H.P.	Richardson Westgarth Double Reduction Geared Turbine, 13,750 H.P.
Unit Capacity: Gal	6,000	6,000
Initial Fill Date:	August 1966	August 1966
Make-up Oil Since Last Sample: Gal	260	145
Date Sample Taken:	5-23-71	8-3-71
RTL Reference No.	R-3,847[a]	R-5,380[a]
Inspection:		
Gravity: $^{\circ}$API	28.2	28.0
Viscosity, SUV: Sec.		
100°F	451	461
130	207	207
210	61.3	61.5
Viscosity Index	99	97
Interfacial Tension, 77°F:Dynes/Cm		
ASTM D 971-50	23	20
Flash, OC:$^{\circ}$F	505	495
Fire, OC:$^{\circ}$F	550	555
Pour: $^{\circ}$F	+5	+5
Color, ASTM D 1500-58T	7.5	8.0
Appearance After Aging 24 Hr, 70°-80°F	hazy	bright
Water: PPM	320	42
Water by Dist'n: %		
ASTM D 95-46	trace	nil
Water and Sediment: %	trace	nil
Carbon Residue, Conradson: %	0.15	0.18
Rust-Preventive Test, ASTM D 665-54		
Procedure A, 24 Hr	passes	passes
Centrifuge Test Separation: %	trace	trace
Neutralization Value ASTM D 974-58T		
Total Acid No.	0.27	0.47
Emulsion Test, 180°F Fed. 3201.5		

Distilled Water	37-25-18 (60)	40-38-2 (60)
Foaming Test, ASTM D 892-58T		
Vol. Foam End of 5 Min		
Blowing: Ml		
Sequence 1	5	5
2	10	25
3	5	5
Vol. Foam End of 10 Min		
Settling: Ml		
Sequence 1	0	0
2	0	0
3	0	0
Oxidation Test, ASTM D 943-54		
Time Oxidized: Hr for 2.0		
Acid No.	240	190
Oxidation Stability: Hr	12	6
Agent 046:%	0.52	0.35
Ash (Max. Temp., 1,000°F): %	0.009	0.009
Analysis Spectrographic: PPM		
Aluminum	2	2
Calcium	1	1
Chromium	0.06	0.04
Copper	10	15
Iron	10	20
Lead	9	8
Magnesium	8	8
Manganese	0.6	0.7
Nickel	0.2	0.3
Phosphorus	15	17
Potassium[b]	1	0.9
Silicon	2	2
Sodium[b]	5	6
Tin	2	2
Zinc	6	6

a. Sample obtained at Green Island Terminal.

b. Determined flame photometrically.

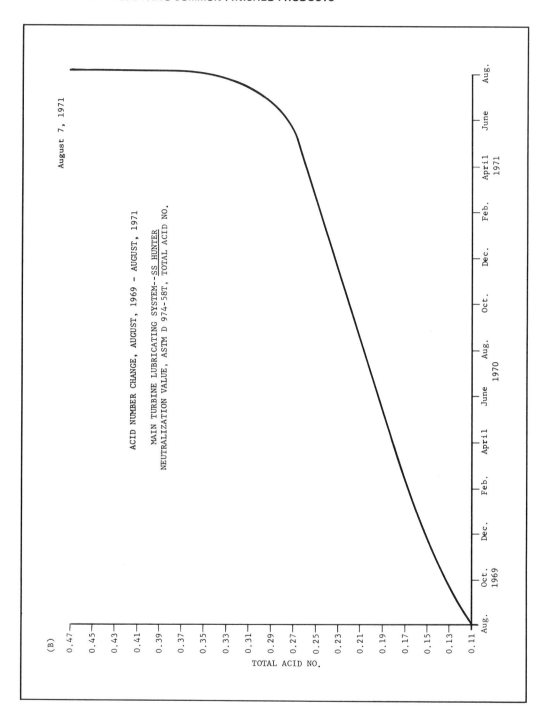

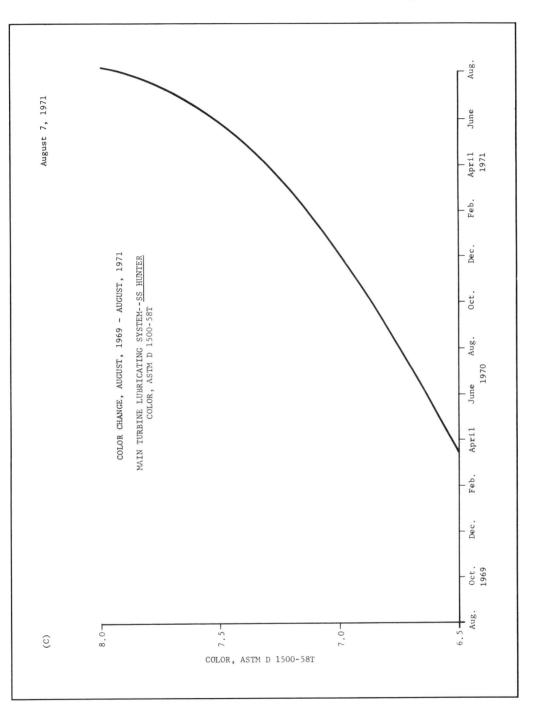

(C)

August 7, 1971

COLOR CHANGE, AUGUST, 1969 – AUGUST, 1971

MAIN TURBINE LUBRICATING SYSTEM--SS HUNTER
COLOR, ASTM D 1500-58T

COLOR, ASTM D 1500-58T

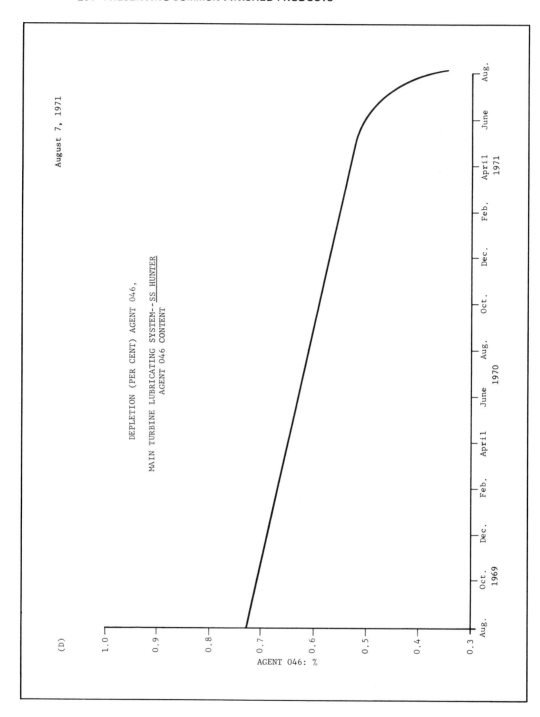

Progress Report

You may also use the outline on pp. 196-97 for writing a report even though the study has not been completed. The only difference is that the results, the discussion of results, the conclusions, and the recommendations are not final. Such a study is called a progress report. The progress report is especially meaningful if there is reason to believe that the final analysis will confirm the tentative results and conclusions. The progress report can also be used to encourage leaders to think more about the problem, or to show them that changes should be made while the investigator is still at work.

Making changes often saves a company a good deal of time and money. It might be a good idea, for example, to change the scope of the study or the methods of collecting data. Perhaps more data should be accumulated, or the scope broadened or further restricted. It may even be decided that the investigation should be abandoned.

From the report writer's point of view, putting the progress on paper should help him to see better the direction in which the results are pointing. Or he may be able to see that changes should be made before the study is continued. Also, of course, he will want to learn management's reaction to what he has accomplished to date.

The following outline shows how the progress report is most commonly structured. Naturally, the outline is in its simplest form; individual sections are not subdivided into parts:

Introduction
Summary of Work Done to Date and Already Reported
Detailed Account of Work Done Since the Last Report
Plan of Work to Be Done Next, According to Schedule
Manpower and Cost Requirements—Past and Future
Recommended Changes in Work to Be Done Later

In a particular project, topical headings—as appropriate—would be used. These in turn would be broken down into subdivisions covered by subheadings. Since the first two major items may have been covered in one or more previous progress reports, the emphasis obviously will be on the last four. Generally, the first two items are included to refresh the reader's memory in continuing investigations.

Like other informal reports, progress reports vary greatly in nature, structure, and length. Also, there may be many progress reports for an extensive project, and as few as one for a shorter one. In most organizations, too, progress reports are written at regular intervals. The interval may coincide with the end of a phase of work, or it may be weekly, bi-weekly, monthly, etc. Leaders' and clients' needs almost always determine how often progress reports are submitted.

Like many other informal reports, the progress report may have to be presented in a formal structure at times. The next chapter shows how any informal report can be presented in formal "dress."

THE IMPORTANCE OF EXPLICIT STATEMENTS OF PURPOSE AND SCOPE

In all of the memos and reports presented in this chapter, you will find explicit statements of purpose and scope. The example memo on page 181 begins with a sentence that covers both purpose and scope:

       We request your approval of our rewriting the specification
    for the metal used to construct centrifugal pump impellers
    that are subject to corrosion in their normal operation.

The specimen beginning on page 192 also states the purpose and scope in the first sentence.

       This procedure can be used to determine nickel in zinc
    and cadmium.

At times, however, you will find it necessary to "put off" the presentation of your statements of purpose and scope together—or to present one of the statements later. For example, the illustration on page 182 begins with a statement covering the scope only:

       For years we have been using the colorimetric method to
    analyze the metals content of catalytic charge stocks. This
    is a very accurate method for determining the amount of
    vanadium and nickel.

The statement of purpose is not presented until the problem has been fully discussed (after the first two paragraphs). It is not until the reader gets to paragraph 3 that he knows what the writer's intention is:

       The advantages of the spectrographic method are obvious.
    We can complete an analysis for vanadium in 5 minutes, and we
    can test samples frequently--daily if necessary. For these
    reasons we think it will be highly advantageous to use the
    spectrographic method instead of the colorimetric.

In this case, you may recall, the statement of purpose—that is, the proposal—was put off because the change might be considered undesirable by leaders.

Statements of purpose and scope also may be put off so that background information can be presented first. The subject may not be controversial at all, of course; it may be only that the writer feels a need to brief the reader—bring him up to date—before he announces his intention and area(s) of concentration. The beginnings of the reports on pages 189-98 and 225, and the proposal on page 190 illustrate this approach. Of these three, we notice, the last mentioned is quite different from the others. The scope here is *very* detailed (seven specific questions are asked), while the statement of purpose—although clearly indicated—is presented "differently," in sentences 2-3 of paragraph 2:

```
      . . . It is obvious to us now that we will need extensive
research on the subject in order to meet the deadline.  Such
a study would include a search for answers. . . .
```

Perhaps the most interesting example of all is the beginning of the proposal on page 184. That specimen has *two* statements of purpose and *two* statements of scope. The first sentence of the proposal states the purpose of the research *project,* and the next three paragraphs cover the *project's* scope. However, for the purpose and scope of the *proposal itself,* we must look to the last sentence under "Purpose and Scope":

```
    This is a proposal for an initial exploration, based on
raw-material, production, and equipment costs previously esti-
mated by the Paint Laboratory.
```

The last example should be kept in mind as we turn to a final point. When you begin a research project, you often seek a "yes" or "no" answer. For example, your problem might be this:

```
    Should we replace the acid-clay contact method with the
hydrotreating process to decolorize lubricating oils?
```

Naturally, at the start of your research, you do not know if the answer is "yes" or "no."[5] When you have completed the project, however, you know the answer. When you write your report, therefore, you have a decision to make: you ask yourself, Shall I state my purpose as

```
    The purpose of this study was to determine if we should
substitute the hydrotreating process for the acid-clay contact
method of decolorizing lubricating oils.
```

or shall I state it as

```
    The purpose of this report is to show that the hydrotreat-
ing process is better than the acid-clay contact method of
decolorizing lubricating oils, and thus should replace it.
```

The answer to your question depends, most probably, on the policy of your organization. However, if there is no established policy, you should use the first statement if the subject is controversial; the second statement if it is not. The first statement is neutral; it is not a commitment. The second statement is a judgment; it *is* a commitment. The first statement indicates that an objective development will follow up to the conclusions section of the report, where the

---

[5] Or a qualified *yes* or qualified *no* (yes if—or no if—the situation is such and such, the application is this or that, etc.).

judgment (hydrotreating is better and should be used) will be stated. The second statement indicates that the development will be focused *throughout* on supporting the judgment stated at the very beginning.

You need to know these two approaches because you may write a number of reports like this one during your career. Commonly known as *feasibility* reports, these presentations help leaders decide if something is or is not so, if something should or should not be done, etc. The approach you use may greatly influence the reader's reaction. If you use the first (neutral) approach, your report—like a mystery story—cannot supply the solution to the end. (You might call this development a "whodunit" one.) If you use the second (judgment) approach, your report must support the solution from beginning to end. (You might call this development a "defense of a judgment," or "justification report.")

Regardless of the situation you find yourself in at the beginning of a research project, remember that explicit statements of purpose and scope are always needed *somewhere* in the beginning. Your reader must always know why you are writing and what, exactly, you are writing about.

EXERCISES

MEMOS

1. Discuss the effectiveness or ineffectiveness of the following informal proposal to make a change.

Date: July 17, 19__

From: Henry B. Allworth

To: Mr. Leonard G. Hahn

Subject: Piling Settlement

A recent investigation of the settlement of the new laboratory annex indicated that some modification of our existing pile-driving specifications is in order.

Presently we are driving our pilings in accordance with the ENR formula. Thus we stop driving a 20-ton piling when the penetration rate reaches 1 inch every two blows with a No. 1 hammer.

In order to eliminate a recurrence of pile settlement, we wish to recommend the following:

1. Make enough test borings to obtain a good picture of the underground strata.

2. Revise the pile-driving specification as follows for 20-ton piling:

> a. A minimum of 55 blows with a No. 1 hammer for the final 3 feet of penetration.
>
> b. Strike a minimum of 30 blows for the last foot of penetration.
>
> c. To control tip brooming, striking a maximum of 40 blows for the last foot of penetration.

We believe that if the above procedure is followed, the amount of pile settlement will be insignificant in the future.

2. Discuss the effectiveness or ineffectiveness of the following informal proposal to carry out research.

Date: February 3, 19__

From: Roger E. Muncy

To: Mr. Carl V. Peterson, Manager of Operations

Subject: Method of Caustic Dehydration in Open Pots

Our present method of caustic dehydration in open pots is both antiquated and dangerous. The present dehydration method involves unnecessary handling of the caustic in small quantities. For each of the open pots, a separate pump is used to pump the caustic to and from the fusion pots. Operating and maintenance costs for the pumps, pots, and piping are high because of the unnecessary handling.

If our proposal is approved, we will seek answers to the following questions concerning suitability of design, economics, additional equipment, and what safety precautions will be necessary for men working in the area if a continuous evaporator is used.

The proposed study should provide a first step toward a better caustic dehydration process.

3. Discuss the effectiveness or ineffectiveness of the following memo of procedures.

Date: August 19, 19__

From: Donald C. Williams

To: Mr. Ralph S. Burns, Owner, Green Valley Farms

Subject: Using Fertilizers Containing Suspensions

This memo describes the procedure for processing, applying, and storage of fertilizers containing suspensions.

Liquid fertilizers are rapidly beginning to compete with bulk (dry) fertilizers. Suspension fertilizer is a type of liquid fertilizer, but it is very different from regular clear-liquid mixtures. Suspension fertilizer is made in much the same way as other liquids, but with little regard to purity and solubility.

Production of Suspension Fertilizer

The basic ingredients for the mixture are water, attagel 150 (clay substance), wet-process phosphoric acid, aqua ammonia, UAN solution, and muriate of potash. Mix the ingredients together in huge tanks at a high speed.

Make a base solution from the ingredients, and add others for different types of blends. Add attagel 150, which is a clay substance made from a colloidal mineral, to stabilize the solution. Agitate the mixture so that heat is evolved to dissolve the salt additives. When the mixture cools, the salts crystallize and the solubility level is reached. The attagel 150 allows the salt to crystallize on it, causing the suspension to hold.

Wet-process phosphoric acid can be used as the source of phosphorus. Iron and aluminum salts settle out of the solution, an unimportant point because purity is not required. When required, potash can be added to the mixture because it can settle out somewhat.

Applying Suspensions

Apply the liquid fertilizer by using either a truck or a tractor with trailer apparatus, the tractor being preferred. The equipment needed to apply the fertilizer is simple but expensive. Two stainless-steel 500-gallon tanks are used, mounted on a skeleton trailer. Each tank should have a small pump with agitator blades in the bottom. These are needed to recycle the suspension constantly so that the fertilizer will remain thoroughly mixed.

Because the fertilizer contains solid material, use special filters in the pipes running to the nozzles. Be sure that the filters are large enough to allow the suspension to escape.

INFORMAL REPORTS

4. Discuss the effectiveness or ineffectiveness of the following informal report.

Date: November 9, 19__

From: K. L. Henderson

To: Mr. P. B. Larsen, Manager of Maintenance

Subject: Water Pump Gear Failure

On November 2, a pinion gear failed in the cooling water pump of No. 51 still. The unit was opened and inspected, and the pinion gear was found to have three broken teeth. At the time it was also noted that the gear spray oil pipe was broken, and it is felt that this may have contributed to the gear failure. This pipe is located at the junction where the pinion gear meshes with the bull gear. Obviously, the spray tube dropped, came in contact with the gear, and caused the failure.

The broken teeth on the pinion gear were removed and the pinion and gear were dressed by hand. As a result of the failure and changing the tooth profile, it was found that when the unit was again placed in operation November 5, the reduction gear was exceptionally noisy. On inspection we found that the gear was vibrating too much for continuous service.

In addition to the gear work, a complete set of new bearings was installed in the reduction gear. The turbine bearings and the pump bearings were also inspected, and all were found to be in satisfactory condition. Finally, we inspected the high and low speed couplings and found these to be satisfactory.

The water pump has been back in operation since November 6.

5. Discuss the effectiveness or ineffectiveness of the following progress report.

Date: December 1, 19__

From: Frank F. Coburn

To: Mr. John T. Perkins

Subject: Acid Plant Pipe Failures--Progress Report

Introduction

The failures of cast iron pipe in the cooling system of the acid plant have become a hazard to the operating personnel.

Summary of Last Report

Preliminary inspection of the increasing number of pipe failures showed that the failures occurred in four forms:

1. Pin hole.

2. Local corrosion at the threaded flanges was noted.

3. Longitudinal splits.

4. There was also corrosion along the cold shuts.

We carefully examined twelve cast-iron pipe failures and recorded the number and types of failures. We had been told that the longitudinal-split failure occurred more frequently than the other types, but examination did not prove this claim to be true. Because of the costs of transporting the heavy cast-iron pipe and sawing it into sections, we selected only two failures for microscopic examination, a 180° return and a piece of 8-inch-diameter pipe 20 feet long.

Present Work

We have completed the microscopic examination of the two pipe samples and have analyzed the results.

Future Work

All that remains to be done is writing the final report on our findings.

Forecast

We expect to complete the final report by December 29.

THE IMPORTANCE OF EXPLICIT STATEMENTS OF PURPOSE AND SCOPE

6. Discuss the effectiveness or ineffectiveness of the statements of purpose and scope in the memos and reports quoted in Exercises 1–5. Rewrite the statements that you think are ineffective, making any assumptions that you think are necessary in each case.

SPECIAL PROBLEMS

7. Write a proposal to make a change.
8. Write a proposal to carry out research, similar to the proposal on page 190.
9. Write a transmittal for a report that you have written, or are now writing, for a course assignment. Address the transmittal to your instructor.

10. Write a set of procedures for laboratory experimentation or a similar series of actions.
11. Write a report on a meeting you have recently attended. Prefer summarizing (what speakers said, what actions were taken, what motions were made, discussed, and voted on, etc.) to reporting every detail, but include details where necessary.
12. Write an informational report on an experience you have had or on some aspect of a research project you are now, or have been, engaged in.
13. Write an interpretive report on an experience you have had or on some aspect of a research project you are now, or have been, engaged in.
14. Write a progress report on a current research project you are working on.
15. Write a report in which you present and discuss final statements of purpose and scope for a research project you have completed or are now completing.

Chapter Ten

FORMAL REPORTS
AND PROPOSALS

The formal report or proposal is not essentially different from the informal in organization, style, or language. It is not necessarily different in length, although usually it is longer than the informal report or proposal. What does distinguish the formal report or proposal is its physical presentation. The paper may be better; the typing may be more accurate. A title page and other prefatory elements usually precede the first page of text. And for obvious differentiation, the whole report is placed in a binder.

REASONS FOR FORMAL PRESENTATIONS

The most common reasons for preparing a formal report or proposal are as follow:
1. The presentation is the only proof that a study (perhaps an extensive and costly one) has been done.
2. From leaders' and clients' points of view, a formal presentation makes a better impression on decision makers.
3. The presentation covers a definite, long period of work (month, quarter, half-year). Therefore, it needs to be distinguished from everyday reports and proposals on the subject.
4. The report or proposal is to serve as a permanent record.
5. An investigation that began as a routine study developed into a more involved one, whose results were found to be more significant than an informal report or proposal would indicate.

Of course, more than one of these reasons may apply.

Whether a report or proposal is formal or informal its purpose is the same. The report concludes a study or a phase of a study; the proposal looks ahead to one or the other.

214

THE FORMAL REPORT

Typical Contents

We may begin our study of the formal report by listing its parts:
1. Cover
2. Title page
3. Table of contents
4. Memo of instructions (Authorization) [1]
5. Memo of transmittal
6. Abstract or Summary
7. Introduction
8. Report proper
9. Summary of conclusions ⎫
10. Recommendations ⎬ Included only in interpretive reports
11. Appendixes ⎭

We can classify the parts of the formal report and analyze them as
1. Prefatory Pages (parts 1–5)
2. Abstract or Summary (part 6)
3. Report Itself (parts 7–10)
4. Appendixes (part 11)

Prefatory Pages

Cover Whether it is long or short, every formal report should have a cover. Often the cover is an inexpensive paper binder, but attention is given to its eye appeal, especially if it is prepared for a client. The minimum information required on the front cover is the title of the report, the author's name, and the date. Frequently added—and prominently displayed—is the company's name.

Title Page As the illustration on page 221 shows, the title page contains more information than the cover. In addition to the title, author's name, and date are the following:
1. The author's title and department.
2. The name and title of the officer for whom the report was prepared.
3. The signature(s) of the author's immediate superior(s), indicating approval of the report and clearing it for submittal to the higher authority. (In the specimen only one signature is required because the writer is a supervisor directly below the manager of operations.) Sometimes the signature of an immediate supervisor as well as that of the head of the department may be necessary.
4. The name, city, and state of the company or company division. This in-

---

[1] Instructions (authorization) from a client, and the transmittal to a client, would be presented in letter form.

formation is helpful when the report is prepared at one division of a multi-plant or multi-office company. (It also helps those, in the company and elsewhere, who cite the report in bibliographies of reports they write themselves.)

Good titles require no little thought, by the way. Good titles always indicate the exact purpose of the report. These titles from formal reports illustrate:

```
"Progress Report on Improving the Quality of Grayson Mills'
   Woolen Textiles" (This could have been written: "Improv-
   ing the Quality of Grayson Mills' Woolen Textiles: Prog-
   ress Report.")

"Feasibility of Producing Cadmium Oxide at the Monmouth
   Plant"

"Eliminating Adverse Effects of Condenser Maintenance:
   Final Report"
```

As these examples also indicate, a title should be clear and completely informative.

Table of Contents   The table of contents in the specimen report (page 222) is typical. Note that text headings are presented in sequence and according to their degree of importance: first-degree headings at the far left; second-degree headings, indented; third-degree headings, indented even farther. Of course, headings of a degree must always "total" the larger heading over them. (See pages 32–33.)

Note also that figures and tables are listed separately, and that each illustration is identified by its number, title, and page.

Memo of Instructions   The memo of instructions (or of authorization, as it is sometimes called) assigns a worker to carry out a study. Ideally, it follows this outline:

1. Define the problem and state the purpose to be fulfilled.
2. Break down the problem into the questions to be answered or the factors to be investigated.
3. Suggest possible sources of information, and indicate how the problem might be attacked, what research methods might be used, etc.
4. Include a deadline for completion of the study and presentation of the report.
5. Note any limitations that have not been covered under item 2, above. (Usually, such limitations are of manpower to be used and money to be spent.)
6. Indicate how the results will (may) be used.

In the specimen report on page 223, note that the memo written by Billings' boss is only an approval of the study Billings proposed (see page 190). Naturally, if a leader—rather than a worker—initiates the study of a problem, the memo of instructions will have to be quite complete. In either situation, it is a good idea to include the memo in the report—to show that the research completed was properly authorized by a responsible officer of the company.

Memo of Transmittal The memo[2] of transmittal (or covering memo, as it is often called) is the writer's personal introduction of the report to his primary reader. The transmittal should do more than merely introduce the report, however; it should show one or more of the following:

1. How the report fulfills the expectations of the reader.
2. How successful or unsuccessful the writer thinks the study was.
3. What special problems, if any, were encountered (such as not being able to obtain expected results during experimentation).
4. How helpful participating individuals, groups, and organizations were.
5. What limitations were imposed by the budget allowed for research.
6. What problems were created by the limited amount of time allowed for the study.
7. What personal insights and benefits were gained through the experience.
8. What, in brief, the writer's conclusions and recommendations are.

The example on page 224 touches on items 1, 4, and 8 in the list above.

Abstract or Summary

As indicated in the discussion of summaries on pages 119–24, a complete summary—10 to 20% of the length of the report—is most meaningful. When such an introductory summary is well done, the writer fully satisfies his superior's request to "put it on a page."

In addition to the introductory summary, some technical reports also include an abstract. The abstract in this case may be topical or informative (see page 119), but it is always brief—a true abstract of 5% or less when the report is fairly long. Frequently it is placed between the title page and the table of contents. The specimen report in this chapter needs no abstract as such, because the essential elements of a combined topical and informative abstract are included in the transmittal.

Note that the summary on page 225 follows the introduction and, therefore, includes nothing from it. The reader must read the introduction to completely understand the summary. In another writer's report, the introductory summary might precede the introduction. There, naturally, the summary would include basic information from it. Both developments are common. However, an obvious advantage of the "introduction first" development is that the summary is shorter.

---

[2] Called a *letter* of transmittal in a report to a client.

The Report Itself

The basic elements of a well-developed report *introduction* are a statement of purpose and a statement of scope. These two elements are covered on pages 206-08, but several points need to be stressed, and several added.

The *statement of purpose* clarifies the objective of the report. Two developments of the statement are possible. If the problem examined is not controversial, you may want to indicate that you have found a solution.

```
    The purpose of this report is to show ways to improve
the company's quality-control procedures.
```

If you began with a statement like this, you should make sure that your whole report is developed to support the assertion made.

However, if the subject is controversial, you may prefer writing a neutral statement of purpose.

```
    The purpose of this report is to analyze the company's
quality-control procedures.
```

This statement is objective; it indicates no judgment. Presumably the report following will not present a judgment until the entire problem has been thoroughly analyzed.

Of course, the second statement above may be used even when the subject is noncontroversial. Some companies insist that writers keep judgments out of their introductions.

The *statement of scope* defines the extent of coverage and the limitations, or restrictions. Thus, the statement includes

1. A list of aspects of the subject to be analyzed, factors (sub-problems) that make up the problem, standards to be applied for evaluation, etc.
2. Any special limitations that may apply, imposed by a leader, the writer himself, or circumstances.

The third paragraph of the introduction to the specimen report (see page 225) illustrates *two* different scopes. The first scope covers the evaluation of electrostatic precipitators vs. cloth-bag dust collectors. The second scope covers the specific aspects of the bag collectors that must be analyzed before a decision can be made.

In some reports scope is presented as a list of questions. With this approach, the report is, logically, the answers to these questions—fully discussed and analyzed, of course. The writer of the specimen report could easily have used such a development; you will recall that his proposal (see page 190) had such a list of questions.

Naturally, limitations vary. For example, the writer might indicate that some points, aspects, etc., were not covered because there was not enough time to do so, or funds were limited, or a leader requested that only economic feasibility be studied, or the worker decided to evaluate only two brands or two processes, etc. (Such changes should always be discussed with leaders first, to be sure. And, where appropriate, the approval of leaders should be obtained *before* the study is completed.)

In addition to statements of purpose and scope, a report introduction may have a *definition of terms*. (Definition, and reasons for including it, is covered on pages 96–103.) A section labeled "Definitions," "Meaning of Key Terms," or "Important Terms" is included because

1. The writer knows that some terms are unfamiliar to the reader.
2. The writer assumes that some terms are unfamiliar to the reader.
3. A word is used in a special way (see page 103).

Finally, many introductions end with a section entitled "Methods of Collecting Data" or a similar phrase. Here the writer tells what kind of sources (and perhaps what specific sources) he relied on for data. It should be apparent that such a section should be included only when the writer believes his reader wants to see it. However, the writer is inclined to include it in many reports because it shows that he relied on more than his own resources.

Sometimes in the introduction, sometimes as first elements in the report following, are such sections as "Background," "Equipment and Materials," and "Procedure." The background section often is historical, tracing work done by previous investigators, the history of a problem, etc. The other two—equipment and materials, and procedure—cover exactly what they say. Naturally, they are used only for laboratory reports on experimentation that is likely to be done again.

The report presented on the following pages contains these introductory elements: a short paragraph on the background of the problem, statements of purpose and scope (two each in this case), and details on sources and materials included in the appendixes.

The *report proper* includes all the items or areas indicated by the statement of scope. Informational reports present the complete facts; the only interpretations are those of the authorities whose facts are cited. Interpretive reports present the complete facts, authorities' interpretive comments, and the writer's conclusions (and, where appropriate, his recommendations). Interpretive reports present solutions to problems, defend judgments, evaluate by standards, etc. Tables and figures appear as they are needed in both informational and interpretive reports—always integrated with the text, of course.

The specimen report following begins with description of a process and interpretation of its efficiency. Then the report weighs the advantages and disadvantages of two types of dust collectors, and judges the bag collector to be better. Having established the basic conclusion and recommendation, the writer then presents, with interpretation, factual information about the specific bag-collector system recommended. This development is logical and effective in this case. In another case, a different development may be needed. (For specific guidelines on applying the principles of technical writing in developing effective formal reports, look under "report" in the index.)

The last parts of the report itself are the *summary of conclusions* and the *recommendations.* These two elements, which are covered fully on pages 126-27, are not always needed. They help the reader of a long report, however, because

1. The summary of conclusions covers all the specific, important interpretations made in the report.
2. In this presentation, the conclusions can be stated in the order of their descending importance: most important first, least important last.
3. The summary of specific conclusions should anticipate the specific recommendations that follow (when recommendations are indicated). Thus, the final two elements have a unity and an emphasis that they could not be given elsewhere in the report.
4. The recommendations, parallel to and complementing the conclusions, are presented also in order of importance. Ideally, there is a recommendation for every conclusion listed.

Thus the summary of conclusions and the recommendations, sometimes called the "terminal summary," both remind the reader of what he has read and what the writer has been pointing to, and also give him needed perspective for considering the problem and the writer's solution.

Appendixes

Appendixes are the final part of a report. Here are found end notes (if footnotes are not used) and the bibliography. Here also are all the adjuncts (included-but-not-important data, computations, supplements, etc.) for the curious reader. An appendix is the place for whatever cannot be presented elsewhere in the report.

A SPECIMEN FORMAL REPORT

Although written by a student, the specimen formal report on the following pages could easily have been written by a professional. The student's purpose was to solve a problem at a plant in his home-town area. Thus we can call the presentation an exemplary interpretive report.

Except for the cover and the appendixes beyond the bibliography, the report is complete.

THE FEASIBILITY OF INSTALLING

A CORWIN GLASS-BAG DUST COLLECTOR

AT ACME CEMENT CORPORATION

by

Lawrence F. Billings

Supervisor, Operations Department

Submitted to

Mr. Robert J. Holbrook

President

September 11, 1971

Approved by:

C. V. Allen

Manager of Operations

TABLE OF CONTENTS

AUTHORIZATION   iii
TRANSMITTAL   iv
INTRODUCTION   1
SUMMARY   1
ACME'S PREHEATING PROCESS--AND THE PROBLEM   2
TYPES OF CEMENT-DUST COLLECTORS--ADVANTAGES AND DISADVANTAGES   3
     Electrostatic Precipitators   3
     Cloth-Filter Collectors   3
     Comparison   4
ADVANTAGES OF PURCHASING THE CORWIN COLLECTOR   4
THE CORWIN GLASS-BAG DUST COLLECTOR   4
     Glass-Bag Data   5
     Maintenance   6
     Pattern of Operation   6
          Dust Collection   6
          Continuous Cleaning Cycle   6
          Dust Removal   6
A CORWIN GLASS-BAG DUST COLLECTOR FOR ACME   7
     Collector Size   7
     Accessory Equipment   7
     Installation Time   8
     Cost Estimates   9
          Equipment Cost   9
          Installation Cost   9
          Operating Cost   10
SUMMARY OF CONCLUSIONS   10
RECOMMENDATIONS   10
BIBLIOGRAPHY   11
APPENDIX 1: FLOW DIAGRAM OF ACME PREHEATING PROCESS   12
APPENDIX 2: SUMMARY OF DATA ON CLOTH-BAG MANUFACTURERS
            AND THEIR PRODUCTS   13
APPENDIX 3: CORRESPONDENCE, AUGUST 26-SEPTEMBER 8, 1971   15
APPENDIX 4: MANUFACTURERS' LITERATURE ON CORWIN GLASS-BAG
            DUST COLLECTORS   20

ILLUSTRATIONS

Figure 1   EFFICIENCIES OF CYCLONE SEPARATORS   2
Figure 2   TWELVE-COMPARTMENT BAGHOUSE   5
Figure 3   LOCATION OF PROPOSED COLLECTOR UNIT AT ACME   8
Table 1   ESTIMATES OF TOTAL INSTALLED COSTS   9

Authorization

ACME CEMENT CORPORATION

Intracompany Correspondence

Not a weeper

Date: August 20, 1971

From: Robert J. Holbrook

To: Mr. Lawrence F. Billings

Subject: Your Proposal on Cement-Dust Control

     Your memo of August 18, to Mr. C. V. Allen, is greatly appreciated. We discussed this morning your fine efforts to reduce the amount of cement dust escaping from the smokestack, with the conclusion that we must look outside the plant for a solution to our problem. You have our full support to carry out the study you propose.

     The questions in your memo point to the answers that we too believe are needed. If your report answers these questions, we should be able to act in time to meet the January 1 deadline for complying with the specification of the new dust-control ordinance.

     I confess that I know little about dust control, Larry. However, I have heard a great deal lately about the virtues of fiberglass bags in comparison with electrostatic precipitators. I am especially interested in the bags made by the Corwin Company; these, according to journal articles, appear to be very efficient.

     The articles in question should be available in the periodicals file of the plant library. May I suggest that you begin your search there?

     Because of the need to solve our problem soon, it will be appreciated if your report is completed by September 15. The budget for this project is "open"--within reason, of course-- and I personally will be glad to help in any way I can.

iii

Transmittal

ACME CEMENT CORPORATION

Intracompany Correspondence

Date: September 11, 1971

From: Lawrence F. Billings, Supervisor, Operations Department

To: Mr. Robert J. Holbrook, President

Subject: Report on Solving the Dust-Control Problem

Attached is my report on the investigation carried out to solve our dust-control problem. Answers to all the questions raised are answered in the summary on page 1, and complete details are provided in the report.

The report covers first the nature of our specific problem with dust control. Then it presents the advantages and disadvantages of the two most popular cement-dust collectors. After ruling out the use of electrostatic precipitators at our plant, the report analyzes in detail the operation, installation, and costs of the system using the Corwin glass-bag collector. This system is strongly recommended.

I am indebted to Mr. Allen for his help in analyzing the preheater process and pinpointing the problem. Also very helpful were Mr. Grant and Mr. Vale of the Corwin Company, who supplied general information and specific data on the installation of a Corwin collector system at Acme.

Although pressed for time, I believe that this report is complete in every respect. I am confident that implementing the recommendation will solve our problem.

If there is anything else I can do, please call on me.

iv

THE FEASIBILITY OF INSTALLING A CORWIN GLASS-BAG DUST COLLECTOR AT
ACME CEMENT CORPORATION

INTRODUCTION

In the past few years, public awareness of air pollution in the Valley
has been increasing rapidly. As a result, the County early this summer
passed a dust-control ordinance that is to become effective January 1, 1972.
The study covered by this report was undertaken to enable Acme to comply
with the ordinance, which specifies a limit of .5 lb of dust per 1,000 lb
of standard stack gas.

The purpose of this report is twofold:

1. To show objectively that the fiberglass-bag dust collector is
    ʽbetter for Acme than the electrostatic precipitator.

2. To analyze the operation, installation, and costs of the Corwin
    fiberglass-bag collector recommended for Acme.

First, the report covers the advantages and disadvantages of electrostatic precipitators and glass-cloth dust collectors, in turn. Following
is a comparison of their efficiency, maintenance costs, operating costs,
and installed costs. Second, the report analyzes the Corwin fiberglass-
bag collector system and then proceeds to discuss the size, accessory
equipment, installation time, and costs of a Corwin compartment baghouse
for Acme.

Information for this report came from journal articles, publications
of manufacturers of dust-collecting equipment, and interviews and correspondence with Corwin officials.

SUMMARY

Glass-cloth dust collectors are cheaper to install, easier to maintain, and more efficient than electrostatic precipitators. For several
reasons, the unit manufactured by the Corwin Company is recommended over
that of any other bag-system manufacturer.

Corwin collectors are designed for easy access and minimum maintenance. Bags can be replaced quickly, and individual bag compartments can be
cleaned while other compartments continue to function. The dust can be
disposed of or recovered and recirculated.

For Acme, a Corwin four-compartment, No. 67 unit (20' x 32.5' x 70'
high) will filter the preheater gas with complete efficiency. Only 650
square feet are needed for the installation, and accessory equipment is
available for recovering the dust. Total costs of the unit, installed,
will be about $104,000; annual operating costs will be about $3,520. The
unit can be installed before January 1, with shut-down time of less than one
day. Thus, signing a contract at once is highly recommended.

1

ACME'S PREHEATING PROCESS--AND THE PROBLEM

Acme's preheater contains four cyclones in which the raw material is heated to about 1,500°F before entering the rotary kiln for processing. The raw material is dropped through the four cyclones onto hot gases coming from the rotary kiln.

The cyclones are conical-shaped mechanical separators which admit the gas near the top and allow solid particles to fall out at the bottom. The raw material, which enters at Stage I of the process, falls through each stage below, successively, and finally enters the kiln through a feed pipe from the bottom of Stage IV. The hot gases from the kiln start at Stage IV and pass up through each stage. A flow diagram of the complete process is shown in Appendix 2 (page 13).

The gases leaving Stage I have cooled to about 500°F and have picked up about 4% of the material fed to the preheater. The dust-laden gases pass through a duct and enter a battery of six cyclone dust collectors. These collectors remove only about 50% of the dust in the gas, because about half of the dust is in the 0-20 micron range. Figure 1 illustrates the low efficiency of cyclone separators when the particles are smaller than 20 microns. As the gas is passed through an exhaust fan and out

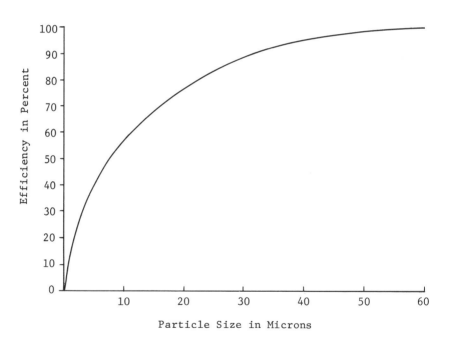

THE EFFICIENCIES OF CYCLONE SEPARATORS

Figure 1

SOURCE: S. Levine, "What You Should Know About Dust Collectors," Rock Products, vol. 68, February 1965, p. 61.

through the stack, it now contains about 2% of the material fed. Therein lies our problem: this amount is much too large to meet the specifications of the County's ordinance.

TYPES OF CEMENT-DUST COLLECTORS--ADVANTAGES AND DISADVANTAGES

Electrostatic Precipitators

Electrostatic precipitators operate at high voltages. The voltages electrically charge the dust particles, which are then collected on electrodes having an opposite charge. The dust is removed by a mechanical rapper which strikes the electrodes rapidly. The dust falls into hoppers below, and is either recirculated or trucked to disposal sites.

Efficiencies range from 97 to 99%, depending on how well the dust is ionized. The electrostatic precipitator works best on wet-process material because the moisture ensures proper ionization. The precipitator operates at a very low pressure drop of about .5 in. of water.[1]

Electrostatic precipitators are used in the cement industry, but three major problems have arisen during their operation:

1. Arcing of the electrodes. Arcing negates the charge on the dust and results in a loss of efficiency.

2. Need for a rectification power source which requires an exceptionally clean atmosphere and special maintenance.

3. Extreme sensitivity to the gas-flow rate, because a definite exposure time is needed to ionize the dust. Volume surges often send dust particles past the electrodes and out through the stack.[2]

Cloth-Filter Collectors

In the cloth-filter collectors, dust-laden gases are drawn through tubular or cylindrical cloth filters because of a pressure difference across the collectors. If silicone-treated glass bags are used, temperatures up to 550°F can be tolerated. Efficiencies of 99 to 99.99% are achieved, and cloth filters have a long life and require no special maintenance.[3]

The operating cost of a cloth-bag collector is high because of a large pressure loss (4 to 10 in. of water) attributed to the resistance of the filter. This resistance requires larger exhaust fans--and therefore there is a higher consumption of power.[4]

---

1 R. J. Wright, "Select Carefully--Dust Collectors Fit Different Needs," Plant Engineering, vol. 19, June 1965, p. 121.
2 Ibid, p. 123.
3 "Giant Portland Cuts Costs, Ups Efficiency with Glass-Bag Dust Collector," Rock Products, vol. 68, February 1965, p. 91.
4 J. H. Bergstrom, "Panel Probes Dust Collection Problems," Rock Products, vol. 68, February 1965, p. 80.

3

<u>Comparison</u>

Generally, cloth-bag collectors are more efficient than electrostatic precipitators. They are considerably more efficient at plants like ours where dry-process material is used. Although fiberglass-bag collectors consume more power, they cost less to maintain. Indeed, the operating costs of the two systems are essentially the same.[5] Furthermore, for a flow rate of 44,000 cfm, the cost of the cloth-bag collector, installed, is approximately 30¢ less per 10,000 cfm than that of the electrostatic precipitator.[6]

<u>ADVANTAGES OF PURCHASING THE CORWIN COLLECTOR</u>

Of the several cloth-bag dust collectors manufactured,[7] the unit made by the Corwin Company is best for Acme. Although the Corwin unit is of no better quality and is no less expensive than others on the market, it has three major advantages:

1. The company is located in Coopertown, less than 50 miles away. This fact will favorably influence the total costs of installation.

2. Engineering services will be supplied promptly if an emergency should ever arise.

3. The Corwin system provides for continual cleaning of the individual compartments while the unit continues to function.

<u>THE CORWIN GLASS-BAG DUST COLLECTOR</u>

The Corwin glass-bag dust collector consists of a plate-steel baghouse divided into equal-size compartments. There are two walkways in each compartment permitting easy access to the bags. The bags are suspended vertically in rows on each side of the walkways.

At the bottom of each compartment is a dust hopper. Each hopper is equipped with an air-locked rotary feeder which discharges the dust without disturbing the pressure in the collector. Each compartment has an air valve which stops and starts the flow of gas. The valve has a relief damper which admits the counter-flow of air for cleaning the bags.[8] Figure 2 on the next page shows side and end views of a 12-compartment baghouse.

---

[5] Wright, <u>loc. cit.</u>
[6] <u>Ibid.</u>
[7] See Appendix 2, "Summary of Data on Cloth-Bag Manufacturers and Their Products" (page 13).
[8] <u>End Your Plant's Dust Problems</u>, Coopertown, Pa.: The Corwin Company, 1970, p. 3.

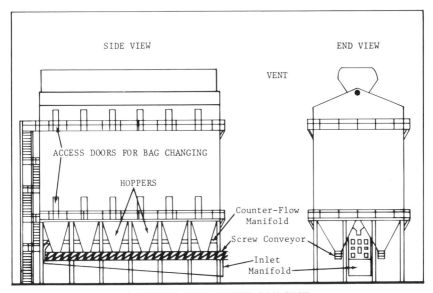

A TWELVE-COMPARTMENT BAGHOUSE

Figure 2

SOURCE: Based on End Your Plant's Dust Problems, Coopertown, Pa.:
The Corwin Company, 1970, p. 3.

Glass-Bag Data

Standard glass-cloth bags are 11¾ in. in diameter and 30 ft long.
They hang vertically from chain hooks and are connected to floor-plate
collars by quick-release clamps. Silicone-treated glass bags can tolerate
gas temperatures of 550°F.[9]

The life of the bag depends on several factors:

1. Temperature of the gas.

2. Abrasiveness of the dust.

3. Tension on the bags.

4. Position in the baghouse.

The life of bags varies even within the same compartment. On the average,
however, they last between one and two years.[10]

---

9   S. Levine, "What You Should Know About Dust Collectors," Rock Products,
vol. 68, February 1965, p. 56.

10 Interview with Mr. G. E. Vale and Mr. B. L. Grant, The Corwin Company,
Coopertown, September 2, 1971.

Maintenance

The Corwin collector is easy to maintain. Besides periodic lubrication of the moving parts, the only routine maintenance needed is changing the bags. The Corwin method of bag suspension helps to keep bag-changing time at a minimum. The bags are held in place by hooks at the top and quick-release clamps at the bottom. Thus, bag replacement is fast and easy.[11]

Pattern of Operation

Dust Collection

Dusty air is blown through a duct by an exhaust fan on the inlet side of the collector. The air enters the hopper of each compartment, where a baffle forces most of the heavier dust into the hopper bottom. The finer dust passes up into the glass bags and is trapped by the pores of the fabric. Clean air passes through the bags and is exhausted to the atmosphere.[12]

Continuous Cleaning Cycle

Since glass bags are too fragile to be shaken,[13] dust must be dislodged from the fabric by other methods. The most common method for dislodging cement dust is reversing the air flow through the bag. The counter-flow of air dislodges the dust, which falls into the hopper below.

For most dust-collection processes in the cement industry, a continuous cleaning cycle is desirable. In the Corwin collector, one compartment can be shut off for cleaning while the other compartments continue to filter. A Corwin program timer automatically controls the cycle as follows:[14]

1. The air valve on one compartment closes, and a relief damper admits counter-flow air to clean the bags.

2. Counter-flow air cleans the bags for a predetermined period of time.

3. The relief damper closes, the air valve opens, and the compartment begins filtering again.

Dust Removal

Dust collects in the bottom of the hopper. It is discharged from the hopper intermittently by the air-locked rotary feeder. The feeder can either discharge the dust into a conveyor system for recovery or into trucks for disposal.

---

[11] *End Your Plant's Dust Problems*, p. 7.
[12] *Automatic Filters*, Coopertown, Pa.: The Corwin Company, 1971, p. 5.
[13] Interview with Mr. G. E. Vale, The Corwin Company, Coopertown, September 2, 1971.
[14] *End Your Plant's Dust Problems*, p. 3.

A CORWIN GLASS-BAG DUST COLLECTOR FOR ACME

Collector Size

The size of a Corwin glass-bag collector is determined by the gas-flow rate and the air-to-cloth ratio. The air-to-cloth ratio is a ratio of the gas-flow rate in cfm to the area of the cloth in sq ft. The recommended air-to-cloth ratio for counter-flow air cleaning is about 2:1.[15]

The net cloth area of a collector is the area that is filtering while one compartment is out of service for cleaning. The net cloth area needed is expressed as a mathematical relation of the gas-flow rate and the air-to-cloth ratio:[16]

$$\frac{\text{flow rate}}{\text{air:cloth ratio}} = \text{net area needed}$$

Since the average flow rate of the preheater gas at Acme is 44,000 cfm, about 22,000 sq ft of cloth area is needed. A standard Corwin unit is available with a net cloth area of 21,819 sq ft. The unit has four compartments, each containing 78 bags. As the standard size of this unit is 20 ft wide and 32.5 ft long, it occupies a ground area of 650 sq ft. Figure 3 on page 8 shows where the unit whould be placed; there is more than enough area available to install the unit in this position.

Accessory Equipment

Although the basic Corwin units are standard, a number of accessories are available to satisfy different companies' needs. The type of accessory equipment required is determined by

1. What size dust-collector unit is needed.

2. Whether a continuous or intermittent cleaning cycle is desired.

3. Whether the dust is to be recovered or disposed of.

Since most of our present plant processes are automated, a continuous automatic cleaning cycle would be desirable. For Acme, it will be feasible and economical to recover the dust. Because screw conveyors are used extensively in the plant, a screw conveyor would be appropriate for moving the dust to the storage bins. Figure 2 on page 5 shows a conveyor in position for a 12-compartment baghouse.

---

[15] Correspondence from Mr. B. L. Grant, The Corwin Company, dated September 8, 1971.
[16] Determining Collector Size for Your Plant, Coopertown, Pa.: The Corwin Company, 1971, p. 12.

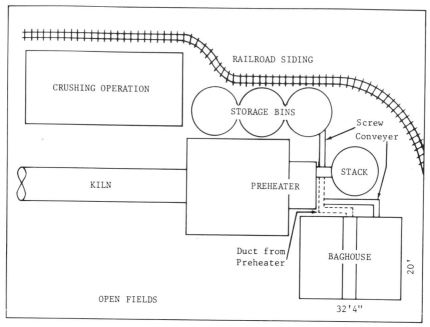

PRESENT STRUCTURES AT ACME AND PROPOSED LOCATION OF THE BAGHOUSE.
Figure 3

For a four-compartment No. 67 glass-bag dust collector with a continuous cleaning cycle and a screw-conveyor system, the following accessories will be needed:[17]

1. Four air-operated butterfly-type air valves.

2. Four 10-in. rotary feeders.

3. One No. 550 I.E. Fan (109.2 hp).

4. Approximately 100 ft of 9 in. screw conveyor.

Installation Time

Because each installation has unique features, an accurate estimate of the time required for installing a No. 67 collector cannot be made at this time. However, Corwin has assured us that the unit can be installed and in operation by January 1, 1972.[18]

---

[17] Correspondence from Mr. B. L. Grant, The Corwin Company, dated September 8, 1971.

[18] Interview with Mr. G. E. Vale, The Corwin Company, Coopertown, September 2, 1971.

Actual construction work can go on without affecting the operation of our preheater. A maximum shut-down time of two working shifts (16 hours) will be needed to connect the dust collector to the preheater.[19]

Cost Estimates

Table 1 presents an itemized account of total installed costs. The approximate total cost of the total installation, including equipment, is estimated to be $104,000.[20]

Table 1

ESTIMATES OF TOTAL INSTALLED COSTS

Item	Cost
Corwin No. 67 four-compartment baghouse	$65,600
Four air-operated butterfly-type air valves @ $600	2,400
Four 10-in. rotary feeders @ $550	2,200
One No. 550 I.E. Fan	4,500
Approximately 100 ft of screw conveyor, 9 in., @ $25/ft	2,500
One Model V program timer	1,500
Construction (total)	25,300
Total	$104,000

Equipment Cost

The cost of the Corwin No. 67 glass-bag dust collector with accessories is estimated at $78,700. This figure includes the cost of delivery and consultant engineers' services.

Installation Cost

The cost of installation is estimated to be $25,300. This estimate includes the cost of[21]

1. Miscellaneous materials (concrete, electrical lines, duct work, etc.).

---

[19] Correspondence from Mr. B. L. Grant, The Corwin Company, dated September 8, 1971.
[20] Ibid.
[21] Ibid.

9

2. General construction.

3. Steel erection.

4. Electrical work.

Operating Cost

The estimate of operating cost is based on costs experienced by other cement companies having similar equipment. This figure is 8¢/cfm per year.[22] Thus, for our flow rate of 44,000 cfm, the estimate is $3,520 per year for total operating cost.

SUMMARY OF CONCLUSIONS

This study has shown that

1. Glass-cloth dust collectors are cheaper to install, easier to maintain, and more efficient than electrostatic precipitators. A glass-cloth dust collector is better for Acme's problem.

2. For several reasons, the unit manufactured by the Corwin Company is better than that made by any other bag-system manufacturer.

3. Corwin collectors are designed for easy access, minimum maintenance, and continuous service.

4. A Corwin No. 67 unit will filter the preheater gas with complete efficiency.

5. There is more than enough space available for proper installation of the Corwin No. 67 unit.

6. Accessory equipment is available for recovering the dust.

7. Total installed costs will be about $104,000; annual operating costs will be about $3,520.

8. The unit can be installed before January 1, with shut-down time of less than one day.

RECOMMENDATIONS

It is therefore recommended that

1. Acme purchase a Corwin No. 67 collector unit with accessory equipment needed to recover the dust for recirculation.

2. Acme contract with the Corwin Company for installation of the No.

---

[22] Bergstrom, op. cit., p. 79.

67 unit, according to an acceptable timetable and price, as soon
as possible.

BIBLIOGRAPHY

1. Automatic Filters. Coopertown, Pa.: The Corwin Company, 1971.

2. Bergstrom, J. H. "Panel Probes Dust Collection Problems,"
     Rock Products, vol. 68, February 1965.

3. Determining Collector Size for Your Plant. Coopertown, Pa.:
     The Corwin Company, 1971.

4. End Your Plant's Dust Problems. Coopertown, Pa.: The Corwin
     Company, 1970.

5. "Giant Portland Cuts Costs, Ups Efficiency with Glass-Bag Dust
     Collector," Rock Products, vol. 68, February 1965.

6. Levine, S. "What You Should Know About Dust Collectors," Rock
     Products, vol. 68, February 1965.

7. Wright, R. J. "Select Carefully--Dust Collectors Fit Different
     Needs," Plant Engineering, vol. 19, June 1965.

INTERVIEW

1. Grant, B. L., Staff Consultant, The Corwin Company, Coopertown,
     Pa., September 2, 1971.

2. Vale, G. E., Vice President, The Corwin Company, Coopertown, Pa.,
     September 2, 1971.

CORRESPONDENCE

1. Grant, B. L., Staff Consultant, The Corwin Company, September 8,
     1971.

THE FORMAL PROPOSAL

The formal technical proposal is an elaborate communication which is written
to prove one's ability to solve a problem. A formal technical proposal also may
be prepared to forestall a problem or to indicate an area of potential benefit.

The analyses and examples on pages 181-91 focus on the development of
simple and complex proposals for *intracompany* situations.[3] The discussion here
focuses on *extracompany* situations—that is, on preparing proposals for clients
and potential clients.

Some experts believe that the formal technical proposal is the greatest chal-

---

[3] See especially the footnote on p. 183 and the specimen proposal on pp. 184–89.

lenge to a technical writer, because he needs extensive knowledge and discreet salesmanship. Essentially, it is an assertion that the proposer can do a job better than any other organization can. To convince a client, the proposer must demonstrate expertise (know-how) and empathy throughout the presentation.

Proposals are the backbone of progress in business, industry, science, and government. Thus, it is understandable that they are unsolicited as well as solicited communications. Continually, companies face major problems that must be solved, opportunities that should be taken advantage of, etc. Sometimes the proposal is informational only, but more often it is interpretive: usually proposers have to supply clients and prospective clients with more than mere facts. Therefore, the following notes emphasize the development of interpretive proposals.

Typical Contents

Proposals vary greatly in structure and length. However, typical contents can be listed, and these can be added to or subtracted from as the occasion demands. Some of the parts will be familiar to you; others will be new and may seem complex. If you keep in mind the importance of empathy throughout, however, the logic of all parts will be apparent. The general outline has these twelve parts:

1. Cover
2. Title page
3. Table of contents
4. Transmittal
5. Introduction
6. Statement of technical approach
7. Plan of attack
8. List of requirements
9. Schedule of accounting for
10. Statement of expected results
11. Summary
12. Appendixes

Like the parts of a formal report, those of a formal technical proposal can be divided into four sections for analysis:

1. Prefatory Pages (parts 1-4)
2. Proposal Itself (parts 5-10)
3. Summary (part 11)
4. Appendixes (part 12)

To make the analysis of these parts meaningful, we shall refer continually to the specimen formal report (pages 221-35) and the situation in which it was developed. As you will see, this particular situation could have been covered by a formal technical proposal from the Corwin Company. Thus, you are asked to imagine that the report was never written.

Prefatory Pages

Of the first three parts—cover, title page, and table of contents—only the cover and title page need changes. These changes are that on both should appear the word "Proposal" and a reference to the client's request for a proposal. The reference may be merely to a letter of inquiry or it may be to the client's project name and number in a more formal situation. (In the more formal situation, the client will send a number of proposers what is called a "request for quotation" or an "invitation to bid." These phrases, followed by the reference, should be included on the cover and title page.)

Also, a change in title may be in order. When the proposal is solicited, the title—or the idea of the title—is usually provided by the client. (The reason is that he requests several proposals, and a single title for all is logical.) However, when the proposal is unsolicited, the proposer should consider including a judgment word that is favorable to himself. For example, rather than use Billings' report title ("The Feasibility of Installing a Corwin Glass-Bag Dust Collector at Acme Cement Corporation"), Corwin likely would use "The Advantages of Installing . . . ," or "The Superiority of a Corwin Glass-Bag . . . ." A word like "feasibility" is neutral; that is, the conclusion reached may be that something *is* feasible or *is not* feasible. Words like "advantages" and "superiority" clearly indicate a judgment favorable to the proposer.

The major change in the prefatory parts is found in the transmittal, or covering letter. The transmittal in the formal technical proposal is a strong sales message. Thus, the proposer should develop it to convince the client that the proposer has the best approach, the best plan of attack, the most reasonable requirements, the most effective schedule for doing the work and reporting on it, and the greatest likelihood of being completely successful. No explicit comparisons with other proposers are made, of course (usually, one proposer does not know what other proposers are being solicited); however, the transmittal from beginning to end should express the confidence that the proposer has.

Proposal Itself

The heart of the proposal consists of the introduction, the statement of technical approach, the plan of attack, the schedule of accounting for, and the statement of expected results.

Introduction When the proposal is solicited, the introduction probably will get to the point quickly. Ordinarily, no long review of the client's problem or situation is needed; the proposer acknowledges the request for his proposal and then, perhaps in the same sentence, indicates his purpose. The important point is that the proposer state as soon as possible that his understanding of the purpose is exactly what the client expects. A breakdown in communication here would be disastrous.

For example, the Corwin Company proposal might begin as follows:

The purpose of this proposal is to show the Acme Cement Corporation the advantages of installing a glass-bag dust collector manufactured by the Corwin Company.

Perhaps then Corwin would include only one sentence to show that it clearly understood the problem:

A Corwin collector will enable Acme to reduce the amount of dust escaping from its stack to less than the .5 limit required by the ordinance effective January 1, 1972.

After stating the purpose, the proposal should indicate generally the scope of its coverage. For example, the Corwin proposal probably would next state:

To accomplish this purpose, the scope of this proposal covers complete details of design, operation, installation, and approximate cost of a Corwin unit designed especially for Acme.

Finally in the "business section" of the introduction, the proposer should indicate what limitations apply, if any. The Corwin proposal, for instance, would realistically add this note:

Cost figures vary with the individual installation, of course. The estimates here are based on Acme's specified flow rate of 44,000 cubic feet per minute, and on the assumption that the installation will be a normal one on solid ground.

The rest of the introduction focuses on general statements of the proposer's expertise in handling a problem or situation of this kind. Like any good introduction, the beginning of the proposal should motivate the client to read on.[4]
Statement of Technical Approach The statement of technical approach indicates how, generally, the proposer intends to go about solving the client's problem. In this section the proposer indicates a course of action that will include a

---

[4] If the proposal is unsolicited, the introduction necessarily will discuss the problem in full. Usually, however, a prospective proposer will send an inquiry to a prospective client in which he indicates very generally the problem or situation the client faces or is assumed to face. For example, upon learning of the County's dust-control ordinance, the Corwin Company might first write a letter beginning as follows:

Inquiries from other companies in your county indicate that a dust-control ordinance will become effective January 1, 1972. The ordinance, as we understand it, specifies that no more than .5 lb of dust be permitted per 1,000 lb of standard stack gas. Like other companies with a cyclone separator, you may be experiencing some difficulty in reducing to the specified limit the amount of dust escaping from your stack.

visit to check the client's specifications in detail, and then the development of a solution to the problem. The Corwin statement of technical approach, for example, would point to an on-site inspection of Acme's preheater and stack. It would also note, in detail, how problems similar to Acme's had been solved in the past. Individual cases, with names and specific data on problems and solutions, would be included. In each instance, Corwin would be sure to note how much the efficiency of filtering the dust was increased. Finally, Corwin would explain the unique features of its collector system—design, operation, and maintenance—as it indicated how it would apply its know-how to Acme's problem.

Plan of Attack The plan of attack is a statement of specific, individual steps the proposer will take if the proposal is accepted. This section may include indications of research to be carried out, of analyses to be accomplished, of designs to be prepared, of blueprints to be drawn, etc. In other words, the plan of attack makes specific and definite the notes included under the statement of technical approach.

The plan of attack in Corwin's proposal would cover, in sequence, a visit to the Acme plant, an analysis of Acme's particular needs, a report written to supply details of installation with exact cost figures, the drawing of designs and blueprints for Acme, the sending of these to Acme for approval, the prefabrication of the baghouse designed for installation next to Acme's preheater and stack, the laying of the footers for the baghouse, and step-by-step details of transportation to the site, erection of the baghouse, and connection to the preheater and stack.

Note that the plan of attack is a pledge as well as a statement of specific actions to be taken in sequence.

List of Requirements Frequently, the proposer must indicate specific requirements that he will have if his proposal is accepted. For example, he may need special equipment, facilities, etc., that only the client can supply. Or, he may need to have the client accomplish some phase of the work (or the client may insist on this in his specifications). Or, the proposer may need materials, equipment, etc., that some third-party supplier will have to provide. The place for these notes on specifications and "counter specifications" is the section titled "list of requirements."

In Corwin's proposal, obviously there is no need for special requirements. Corwin does all the work on a project, including construction of the baghouse. However, if Corwin was unable or if Acme insisted, the job of laying footers,

---

On this assumption, we are writing to inquire if we may submit a formal proposal on the advantages of installing a Corwin glass-bag dust collector at your plant. . . .

If the reply to this inquiry were favorable, Corwin would then quickly prepare a formal technical proposal which would be essentially a solicited proposal in all respects. (Specifications would be acknowledged if Acme included them in its reply to Corwin's inquiry. They would be ignored—for the time being—if Acme did not include them. Corwin would indicate in the proposal the several collector units it manufactures for the range of flow rates Acme might have.)

block, and plates for the baghouse would in this section be assigned specifically to Acme.

The "list of requirements" section may also include cost estimates along with a schedule showing specific dates on which the client is to make payments. If the section is not used for these purposes, a separate section will be needed. In some cases, a completely separate proposal, on costs and payments only, is presented—especially when financial arrangements and estimates are lengthy and complex.

Schedule of Accounting For The schedule of accounting for one's work on a proposal project indicates by the calendar how much time will be devoted to inspection, experimentation, analysis, drafting, construction, transportation, etc. Usually, the schedule is not precise to the day; it may not even be precise to the week. However, here the proposer states what he expects to do, specifically, in what period of time. Often, he will lump activities together—for example, experimentation, analysis, and resultant drafting may be listed as one phase in the schedule. These he will note as requiring so many weeks of the total project. He may even draw a bar chart, showing by individual bars how much time will be spent on each phase of the work.

Included in the schedule section (and shown on the chart, if there is one) will be the time when progress reports will be submitted, along with requests for approval as necessary.

The Corwin proposal will emphasize very firm dates, of course, since the January 1 deadline is only a few months away.

Statement of Expected Results The statement of expected results indicates the extent to which the proposer believes the problem will be solved. As indicated at the end of the informal intracompany research proposal (page 189 in Chapter 9), this statement is often made in terms of percentage. It goes without saying, that any remarks added here will be made with confidence—as will the actual statement of expected results.

The Corwin proposer will have an easy time with this section, since the company's experience with cement-plant installations has been successful. In fact, on the basis of actual results achieved elsewhere, the proposer could well estimate a specific range of "99 to 99.99% efficiency" as what Acme could expect.

Summary

For the convenience of the client, the proposer should include an informative summary at the end of the proposal itself. This summary should be complete, and it should be at least 10% of the length of the actual proposal. (See pages 125–26 for the discussion of complete informative summaries.)

Note that the summary of a formal technical proposal is presented *after* the proposal. This position is different from that recommended for reports that are noncontroversial in nature. The summary is put last here because the proposer wants the client to read his presentation in full; he does *not* want the client to read the summary first—and perhaps make a judgment on that basis alone. Indeed, in laying out a proposal, the writer will do well not to identify the summary in the table of contents, or to use the heading "summary" in the text. Then the reader will come upon it as he reads—and likely will be quite pleased to find it to help him put the proposal in perspective.

Appendixes

In the appendixes the proposer includes computations, data on specific similar work done for other clients, complete names and addresses of satisfied customers, tables and figures that are supplementary to the proposal itself, and—last but not least—testimonials of customers, and professional biographies of key personnel at the proposer's company who will work on the project proposed.

The Corwin proposal probably would include in the appendixes all of these items, including the names and professional biographies of the two officials Billings interviewed and corresponded with: Grant and Vale.

EXERCISES

THE FORMAL REPORT

1. Discuss the effectiveness or ineffectiveness of the following titles of formal interpretive reports.
   a. "Acid Mine Drainage from Abandoned Coal Mines"
   b. "Livestock Diseases"
   c. "Feasibility of Using Lasers"
   d. "Consequences of Pollution"
   e. "Different Materials for Road Surfaces"
2. Discuss the logic and illogic in the organization of an interpretive report, entitled "Improving the Quality of Woolen Textiles at Loomis Mills," as indicated by the following table of contents. The degree of each heading is indicated by indention. See also Exercises 3-6.

Table of Contents

	Page
Summary	1
Introduction	2
Background of Problem	3
Shrink Resistance	4
Table 1	5

```
      Chart 1                              7
      Wrinkle Resistance                   8
      Improvement of Luster               10

      Chart 2                             11

      Improvement of Washability          14

      Table 2                             16

Summary of Conclusions                   17

Recommendations                          18

Bibliography                             19

Table 3                                  20

Table 4                                  21
```

3. Discuss the effectiveness or ineffectiveness of the following summary, which appeared in the report whose table of contents is shown in Exercise 2. See also Exercises 4-6.

Summary

Wool fabrics are generally considered as having low quality with respect to wrinkle resistance especially. This study on methods to improve these factors reveals that the most practical method is interfacial polymerization. However, this method is best only when chlorine is used.

Since no elaborate machinery or expensive materials are required in the use of this method, the installation cost would be low.

4. Discuss the effectiveness or ineffectiveness of the following introduction to the report cited in Exercise 2. See also Exercises 3, 5, and 6.

Introduction

Even with the large increase in population, our sale of wool fabrics has remained relatively constant. On the other hand, the sale of fabrics by manufacturers of synthetics has increased. This increase has occurred because synthetic fabrics have qualities that wool does not have. These qualities are luster, wrinkle resistance, shrink resistance, and washability.

This study is based on library research, data from our laboratory, and information obtained from research centers such as Gunther Textile Research Laboratories.

5. Discuss the effectiveness or ineffectiveness of the following summary of conclusions in the report cited in Exercise 2. See also Exercises 3, 4, and 6.

Summary of Conclusions

The results of this study show that a large number of treatments have been tested, and that there is not any one ideal method for improving all the factors that affect the quality of wool fabrics.

However, the method that appears to be the most efficient of the three evaluated comparatively and in detail is the interfacial polymerization method using chlorine as the add-on material.

This method does not significantly improve the wrinkle resistance or luster. However, no method improves the wrinkle resistance of wool significantly. The method that is best to improve the luster is the "reducing and setting" method.

6. Discuss the effectiveness or ineffectiveness of the following recommendation in the report cited in Exercise 2. See also Exercises 3-5.

Recommendation

Our company should install the proper equipment for using the interfacial polymerization method.

THE FORMAL PROPOSAL

7. On the basis of the information provided in the excerpts quoted in Exercises 2-6, do you think that the writer of "Improving the Quality of Woolen Textiles at Loomis Mills" could develop an effective proposal from the material in his report? Why or why not?
8. Discuss the effectiveness or ineffectiveness of the following introduction to an extracompany proposal, prepared by a firm specializing in consulting engineering, entitled "Proposal for a New Assembly Line in the Generator Department of Star Generator Company."

Introduction

This proposal is the result of studies made to determine how generator production can be modernized and improved before a production line is installed in Star Generator's new plant. As Plant Manager Stein indicated in his letter of May 12, the methods of production in use before the fire were not efficient enough to meet the requirements of an expanding market. This proposal therefore presents a blueprint for a modern production line, with emphasis on three major areas.

In addition to the expertise of our engineering staff, this proposal is based on the ideas and recommendations of Mr. Stein and Mr. Friedman, Foreman of the Generator Department, and on the latest information in the literature generally and Thomas Micro-Catalogs in particular.

8. Discuss the effectiveness or ineffectiveness of the following summary, which appeared at the end of the report cited in Exercise 7.

Summary

This proposal has shown that a linear assembly line with work stations on either side of a moving conveyor belt will greatly reduce the time lost in the past on material handling. Placing the machines in a semi-circle around work areas adjacent to the conveyor belt will save the employees steps and save Star Generator Company money. Bringing the materials to the workers rather than forcing them to leave their work stations, as was done in the old plant, will reduce confusion and inefficiency.

The use of large machinery such as the Mott rotary-table sand-blasting machine and several Baldwin conveyor belt systems will save Star Generator Company a great deal of time that in the past was spent cleaning parts and carrying parts to work tables.

Time and motion studies have shown that the simpler a man's job is, the less likely he is to waste time and make mistakes. As the proposal indicates in detail, everything a man needs to do his job well should be within his immediate reach.

By far the most important idea presented is the point that any job that can feasibly be automated, must be automated in the best interests of the company.

SPECIAL PROBLEMS

9. Prepare a formal report on the research you have been doing to solve a specific problem.
10. Prepare a formal proposal based either directly on the research you have been doing to solve a specific problem, or indirectly on the formal report prepared for Exercise 9.

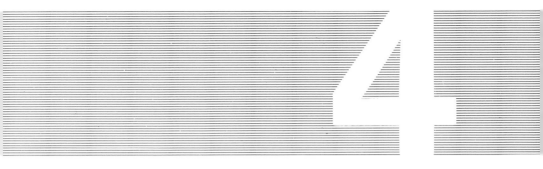

PRESENTING SPECIAL
FINISHED PRODUCTS

Chapter Eleven

ARTICLES

Fellow specialists, workers in other fields, and people in general are greatly affected by scientific and technological progress. Indeed, we might say that, in the final analysis, science and technology exist only for the benefit of man. Yet, for the most part, man knows little of what is being accomplished for his well-being.

Of course, employees in all fields do a tremendous amount of writing in their jobs. Reports, letters, and other communications are prepared regularly, day in and day out. Thus, it may be that the pressure of everyday writing is enough to discourage workers from doing more. However, workers who write articles not only aid other specialists and the public but also benefit themselves.

RECOGNIZING THE BENEFITS OF AUTHORSHIP

The benefits of authorship are cited in a paper written under that title by H. C. McDaniel, for many years Director of Technical Information at Westinghouse Electric Corporation. Addressing Westinghouse engineers at a number of regional seminars, McDaniel said:

> Each article you write benefits you in four ways.
>
> First of all, these articles do much to enhance your professional status. The prestige and recognition you gain from a published article—while less direct—is a more lasting reward than financial compensation. By becoming an author you put your name in front of the engineering profession as an authority on some subject. Not only do your colleagues become more aware of you, but also many outside your field. This brings recognition to you as an individual and often provides one basis for promotion.
>
> Second, this kind of writing trains you in orderly thinking, improves your writing ability, and adds to your storehouse of knowledge—a tremendously important personal benefit that will pay dividends as long as you live. You'll experience the stimulation of authorship, the pride of craftsmanship, and the satisfaction of achievement when you see your name in print; but the really

247

worthwhile benefit from technical writing is the ability you gain to crystal-lize your thoughts and to translate them into words understandable to the reader.

Third is the contribution your published articles make to the advancement of your profession. Lamentable though it may be, the cultural and tech-nological advancement of our entire civilization has been made possible by a relatively small group of dedicated men who were willing to make the sacri-fice in effort and time to record their knowledge. By writing articles, you share your specialized knowledge with your fellow engineers and add to the body of engineering information.

The fourth and final benefit is financial—the money you get for writing technical articles. To be sure, the amount hardly pays you for the time spent researching and writing the article, but it does represent supplemental in-come. . . . When a magazine does not pay you for the article—and a few maga-zines do not—the Company does. So, from one source or the other, you have in the technical article a medium of exchange.

Professional recognition, training in orderly thinking, helping to perpetu-ate the engineering profession, and supplemental income—four good reasons to write.[1]

To pursue the first premise of this chapter, we should note that the readers McDaniel identifies in discussing advancement of the profession ought to in-clude "others who are interested." Also, we should realize that the four benefits are available to *all* workers, not just Westinghouse engineers.

FINDING A SUBJECT

Awareness of the benefits inspires many employees to think about writing arti-cles. However, finding a subject may seem difficult. To help, in *Benefits of Au-thorship* McDaniel lists three categories into which he says most articles fall:
1. Specific applications or installations
2. Solutions to specific problems
3. Review, or "state of the art," articles[2]

The first category, he says, is "one of the most fertile areas" to write about. "Many installations have something unusual or different about them, something that might prove of interest and usefulness to others." To illustrate, he quotes the title of a published article, "Ore Conveyors at the Pend Oreille Mines and Metal Company."

Concerning articles on solutions to specific problems, McDaniel writes: "Many of these are also about applications, but here the emphasis is on a single

---

[1] H. C. McDaniel, *Benefits of Authorship,* Pittsburgh: Westinghouse Electric Corporation, n.d., no consecutive pagination.
[2] *Ibid.*

problem and its solution, whereas the article on a specific application deals with one plant or one application." This problem may be found at more than one company; thus a number of companies may be interested in the solution. As an example he cites "New Electric Brakes Stop Spinning Frames Fast."

Review, or "state of the art," articles "outline present practices or trends in devices, systems, or applications." For illustration, McDaniel quotes such titles as "Timing Devices" and "Electricity in Food Products Manufacture."

Workers who recognize the benefits and begin to get ideas for subjects are well on the way to becoming article writers. Now they can concentrate on whom to write for, how to adapt, and how to write with regularity.

IDENTIFYING READER GROUPS

On the basis of readers' backgrounds and interests, articles are classified as technical, semitechnical, and nontechnical (popular). Primarily, technical articles are written for fellow specialists; semitechnical, for specialists in other fields; and nontechnical, for the general public.

Many publications match these classifications and therefore attract different groups of readers. Physicists read about the latest work on cryogenics in, for instance, the *Journal of Physics*. Microbiologists and other specialists not in physics probably learn what they want to know about these experiments in, say, *Scientific American*. And the general public is satisfied with the journalistic account in a weekly news magazine like *Time* or in a Sunday newspaper supplement. Physicists look for the how-to-do-it article, so that they can duplicate the work, if they wish. Other specialists are receptive to ideas for possible applications in their own research. The public is curious about the effects new experiments on cryogenics may have on technology, society at large, and the individual.

A good way to decide which group to write for is to become familiar with all the publications that accept articles in the field. First, the worker should read the technical journals. (He should have been reading them anyway, of course.) However, he should also become acquainted with less-specialized magazines that publish articles on the subject. At times, there may be no market for a technical article on a certain topic because material on it has recently been published. But at those times, a semitechnical or a nontechnical magazine (or Sunday newspaper supplement) may be seeking an article on that very subject.

When a worker has decided which of the three kinds of publications to write for in a particular situation, he should carefully study the magazines in that class. Every technical and every semitechnical publication caters to a select, limited clientele. Its editors expect that articles submitted for possible publication will be of value to *its* readers. To help authors, therefore, the editors of most magazines publish[3] or mail upon request what is called an editorial state-

---

[3] See, for example, Kirk Polking, ed., *Writer's Market* (published annually), Cincinnati: Writer's Digest.

ment. As the following specimen, the official statement for *Electronic Design,* shows, this information can be extremely helpful to an aspiring writer.

> *Electronic Design* serves the electronic industry—the engineers and engineering management responsible for the design and specification of equipment and devices. Under three broad categories—News, Technology, and Products—it posts its readers on technical trends, gives them practical tips on design, and tells them what products are helpful. The editorial content scans developments in circuitry, microelectronics, test equipment, microwaves, communications, computers, measurements, instrumentation, etc. A new-product section lists a variety of manufacturer offerings, with outstanding characteristics of the products, their price, and their availability.

The editors of *Electronic Design* conclude with the notes that its primary readers are engineering managers and engineers and that its circulation in July 1971 was 70,000.

Reading both the editorial statement and a number of issues of a publication may help a worker see how he might adapt his material to appeal to particular reader groups.

ADAPTING MATERIAL FOR READERS

Many years ago, Gerard Piel, formerly Editor and now Publisher of *Scientific American,* suggested that specialists write understandably and interestingly, so that others—as well as highly specialized colleagues—would be attracted.[4] Yet many technical journals, including some of the more prestigious, continue to publish highly technical articles without concern for their readers. As a result, some specialists are having difficulty reading with understanding—and therefore with interest—articles written by their fellows.[5] Consequently, it is recommended that workers make changes, radical ones if necessary, in on-the-job writing before they submit it as an article. Piel's suggestion makes good sense: obviously, the clearer and more interesting technical writing is, the larger its audience will be.

The plain fact is that readers do not have to read articles as they have to read reports, etc. They can skim articles, read only sections, or even completely ignore them, if they wish. They cannot take these liberties with reports and letters that are addressed to them on the job. Yet, although adaptation is a major challenge, writing articles is an excellent means of developing technical-writing skills further. And making readers glad that they have read an article is an achievement to be proud of.

---

[4]Gerard Piel, "Writing General Science Articles," *Journal of Chemical Education,* January 1954.
[5]See especially F. P. Woodford, "Sounder Thinking Through Clearer Writing," *Science,* May 12, 1967.

The following notes cover adaptation in the four major elements of article structure: title, introduction, body, and ending.

Title

Since the title is what the reader sees first, the writer must be sure that it is clear and interesting. The reader may not bother to go on if he has trouble understanding what the article is about and if the title does not attract his attention.

Two articles cited later in this chapter have titles that are worthy of comment: "Silicones: A Key Ingredient in Aerosol Furniture Polishes" and "Sky Bus Puts Electronics in the Driver's Seat." These are good titles because the first words in each catch the reader's eye, give him an idea at once of the subject, and motivate him to read on.

Titles should be short but not at the expense of completeness. Titles sometimes are shorter than the two just cited, but rarely are they much shorter when subjects are restricted and specific—as in these examples.

Introduction

The same standards that apply to the title are criteria for evaluating the effectiveness of an introduction: be short but complete, get the reader interested, announce the subject, give him reason to read on, and of course be clear.

Dependent on the subject and the writer's point of view, an introduction is developed in a variety of ways. Some of the more common of these are as follow (naturally, two or more of these beginnings may overlap, and one or more may be repeated):

1. analogy
2. cause
3. comparison
4. contrast
5. definition
6. description
7. effect
8. example
9. fact
10. generalization (factual)
11. history
12. interpretation
13. question
14. quotation or paraphrase
15. statement of problem
16. statement of purpose
17. summary

To illustrate overlap and repetition, the following citation—from the introduction to an article on silicones[6]—has interpretation, (erroneous) cause and effect, fact, quotation, contrast, and (overall) definition.

The author begins with the note that aerosol furniture polishes are popular because silicones' "superior gloss and ease of application have made them almost a necessary ingredient." He continues:

---

[6]R. J. Thimineur, "Silicones: A Key Ingredient in Aerosol Furniture Polishes," *Aerosol Age,* September 1970.

And yet, after many years of successful use, there is still some adverse commentary about silicones "scratching" furniture or "lifting" the finishes. . . . For example, one reference by a nationally syndicated columnist suggested that because silicones are based on "silicon, which is derived from sand," [they] therefore may cause scratching or other damage to a furniture finish. In reality, the dimethyl silicone fluids used in polishes are pure polymeric materials completely free from gritty impurities. In addition they are among the best lubricants for wood, plastic, rubber, and many other . . . materials.

Since most articles present more than facts and factual generalizations, the role of interpretation in introductions warrants some discussion. In the excerpt just quoted, four interpretations are apparent: first, a defense of a judgment by the author ("furniture polishes . . . 'necessary ingredient.'"); second, a counter-judgment by critics ("And yet . . . furniture finish"); third, a conclusion by the author ("In reality . . . impurities"); and fourth, an overall evaluation by the author ("In addition . . . materials"). Obviously, interpretation has a vital role when the purpose is to present conclusions, hypotheses, evaluations, judgments, proposals, and recommendations. The writer must always set the stage for his audience.

A final point about introductions concerns length. Although short introductions are recommended, fairly long ones may be needed. For example, the introduction to the article on silicones goes on for several more paragraphs because the author wants to present other claims before he states his purpose: to report the results of tests made to determine if silicones in furniture polish are really harmful.

Body

The body of an informational article includes facts and factual generalizations; the body of an interpretive article includes a full development of interpretations as well.

Here, as appropriate, the author uses the techniques of development that are listed on page 251. Here, also, he includes the adjuncts that reinforce and dramatize the material he presents: tables, charts, graphs, drawings, and photographs.

Finally, if the author wishes to reach more readers than the specialists in his field, he will follow Gerard Piel's advice to include definitions, examples, even explanations of basic principles if they are necessary. Of course, the challenge in attempting to write for more than fellow workers is always to include explanations without obtrusively interrupting the smooth development of essential ideas. Some of the italicized commentary of pages 253–57 illustrates how this challenge might be met.

Ending

The ending of the article should make the reader feel that he has shared a complete experience. Some of the common ways of ending an article are as follow.

If the article is factual, the writer may generalize, cover applications, project into the future, or simply stop when he has developed his last point fully. If the article is interpretive, the author may present his major conclusions, offer recommendations, suggest how his solution might be implemented, or—if he is theorizing—speculate about or analyze the implications of his work.

An illustration of ending interpretive articles is the section "Conclusions" for the article on silicones cited on page 251.

> Our tests showed no adverse effects directly attributable to silicones or a formulated silicone polish. The only adverse effect noted was due to heat cycling, which caused some of the finishes to craze or stress crack. However, there was no pattern to indicate the polished or unpolished surfaces were to blame. A rather random effect was noted.
>
> It can be logically argued that it is impossible to create conditions in the laboratory that correspond to furniture that has been in use for a five-, ten-, or fifteen-year period. However, the laboratory tests were more severe on the finishes than normal use would be. Thus, the problems that arise where silicone polishes are blamed for defects are likely due to inconsiderate treatment of the furniture . . . .

To make the discussion of title, introduction, body, and ending more meaningful, a complete article is quoted and analyzed on the following pages. Comments on specific and general techniques of adaptation appear in italics on the right.

SKY BUS[7] PUTS ELECTRONICS IN THE DRIVER'S SEAT [8]

Clear, complete, catchy title.

Westinghouse electric transit system for cities runs on tracks and is controlled automatically by wayside computers at speeds from a crawl to 50 mph.

Subtitle is an effective one-sentence abstract.

Roger Kenneth Field, News Editor

"Any community that thinks . . . it can assure mobility on schedule for large masses of people without rapid transit . . . is suffering from that 20th-century opiate of the people—the private automobile. A community may buy buses and lay highways until it is blue in the face, but will not have mo-

Quotation beginning sets the stage for the rest of the introduction, paragraphs 2 and 3. Note the two interpretations, the second reinforcing the first, and the two contrasts indicated within them. Note also the in-

---

[7] Although "Sky Bus" is used in this article, Westinghouse itself does not use the name because another company has commercial rights to it.

[8] Roger Kenneth Field, "Sky Bus Puts Electronics in the Driver's Seat," *Electronic Design,* March 15, 1966.

bility until substantial numbers of its public-transit vehicles ride on exclusive rights-of-way."–Leland Hazard, chairman of the Rapid Transit Committee of the Port Authority of Allegheny County, Pa. in his keynote address at the recent First International Conference on Urban Transportation in Pittsburgh.

formation on the speaker's authority.

The managers of some cities have reacted to Hazard's statement by pointing out that manually controlled subway systems are far too expensive for all but the biggest cities. "Design a new subway and make it attractive and cheap," they say.

Transitional paragraph. Note the contrast with the ideas in paragraph 1 and the informal use of quotation.

To see how one electronics manufacturer is responding to this critical transit need, . . . *Electronic Design* visited the experimental track of the Westinghouse Sky Bus in South Park, outside of Pittsburgh.

End of introduction. Statement of purpose carries forward the ideas of paragraphs 1 and 2; sets stage.

I went for a ride around the test track with Ray Fields, head of the Sky Bus project; Don Little, designer of the unique control system, and Dixie Howell, lead project engineer. Fields explained that the Sky Bus is for use by cities with populations as small as 300,000. The trains require no motorman or conductor, and, in fact, the whole system can run perfectly with but a single person monitoring the console of a remote central computer.

Background information, including capsule statements of the potential use and of the basic operation of the system. The use of personalizing is appropriate and effective in this case.

The Sky Bus is powered by two 60-horsepower d-c motors in each car. Small digital computers placed at the wayside control approaching trains and monitor departing trains. Should a computer fail, the computer at an adjacent station on the line controls approaching and departing trains, and service continues uninterrupted.

Details on the system's components and their operation. Note the selection of the major details.

A Smooth, Rubber Ride

The cars have double sets of rubber tires that roll smoothly on concrete tracks. The designer used large thyristors to control the speed of the motors. Acceleration is smooth, yet brisk. As an experienced city subway rider, accustomed to lurches, I was pleas-

Having given the reader an overall view, the writer now presents more-precise details about the parts and their function. Note the explanation of "thyristors" in sentence 2; its gives the reader

antly surprised at the absence of them in the Sky Bus.

The computer at the target station controls the effective dc to the motors by triggering the thyristors during a desired part of each cycle of the three-phase supply current. This current enters the car through a set of brushes that contact the bottoms of the three-phase power rails. The wayside computers can control the train from full speed to a crawl and bring it to a halt within a few inches of its target alongside the platform.

In addition to accelerating the cars, the d-c motors also do most of the braking. Giant open-wound resistors shunt across the armature to brake the train and bring it to very slow speeds. The final braking is done by air brakes because the shunted-motor braking effect decreases as the armature speed decreases. The air brakes, then, are needed to prevent creeping.

In a typical installation one large Westinghouse Prodac 550 computer oversees the small wayside computers, and one person monitors its main console. A loop on the console reproduces in diagram the layout of the track, and lights give the monitor a visual indication of the positions of all trains.

Trains can leave a station every two minutes. The Prodac 550 even decides how many cars each train needs to accommodate passengers waiting on stations up the line. The Westinghouse project team has designed a special set of tracks that enables the computer to direct the coupling of these cars without a single attendant. This is why fine control of the crawl speed is important, Little pointed out.

The cars are coupled both mechanically and electrically. Each couple contains a giant connector. The connector is shielded from the atmospheric elements by a plate, and when one couple presses against another, the plate retracts. Howell pressed the plate to expose the connector.

as much definition as he needs. Note also the personal touch at the end of the paragraph; it gives the reader something of the feel of the riding experience; he likely also knows what a "subway lurch" is like. The further details about motor speed and braking help to sustain the feeling. Finally here, note the clear-cut explanations of the functioning of the d-c motors as main brakes and of the air brakes to do the braking at the very end of the action. Why the d-c motors cannot do all the braking is carefully spelled out.

The reader, having vicariously shared the experience of the ride, is now ready to learn more about the design and operation. Worthy of note is the simplified explanation of the role of the computer. Also effective here is the handling of transition throughout.

The fairly detailed description of the coupling process reveals the writer's interest in even the less-important aspects of operation.

Fields, dressed in a dark pin-striped suit, held a light while we crawled under the bus to get a picture of the unusual antenna that transmits information via a track wire to the nearest wayside computer. A transmitter aboard the train continuously informs the computer of the status of a number of sensors on the cars. These sensors measure train speed, acceleration, card load, wheel slippage, motor overload, air-brake drag, air-brake pressure, and low line voltage.

A simplified explanation of how the system works electrically only hints at its complexity—a tribute to the writer's judicious selection of details.

Fail-Safe Action Provided

Fields said that the designers of the system have programmed the computer to take appropriate action for any failure. It might simply inform the console attendant of minor irregularities, but it would also pull the main power switch and apply all emergency brakes in case of a serious malfunction. Howell stressed another safety feature inherent in the design of the Sky Bus: four pairs of guide wheels, backed with steel plates, keep the cars on the tracks.

Also apparent is the author's anticipation of a key question from readers being informed about an automated system: "What happens if—?"

I found the Sky Bus system aesthetically attractive. The cars, less than one-fourth the weight of standard subway cars, have clean, modern lines. The tracks and their supports do not mar the landscape, and they are not nearly as expensive as standard railroad supports. The Sky Bus can travel either overhead or on or below the ground at speeds up to 50 mph. It is silent and fume-free.

The writer's final remarks about the Westinghouse project answer readers' questions about how good the system is. Note the criteria and the individual evaluations. The author's overall evaluation, a favorable one, is quite comprehensive.

For cities, the cost factor should be important, too. Operation of the transit system requires very few people. And because there are no computers aboard the trains—they just take orders from the wayside computers—cars are not laden with expensive electronic gear.

The train I rode in worked well even with ice on the tracks.

This project demonstrates just one of a number of approaches to the mass-transit problem. Other manufacturers offer ap-

For obvious reasons (Westinghouse is not the only manufacturer of a system for solving the transit problem), the writer ends

proaches that differ considerably from the *with a neutral conclusion. The* Sky Bus. Their designs include alternative *mention of other manufacturers* track material, car size, propulsion systems *probably was intended to give* and track configurations. But all use elec- *the reader hope that the prob-* tronic controls. *lem introduced at the begin-*
ning will be solved one day—
and reasonably.

This article illustrates some of the ways in which important technical information can be made understandable and interesting to a variety of readers. Obviously, the article was written with engineers and engineering managers—the magazine's primary audience—in mind. However, the article would certainly satisfy the informational needs of other readers who were concerned about the problem cited and who might someday have a voice in selecting a system (like the one described here) to solve it.

WRITING ARTICLES REGULARLY

The worker who publishes an article will have good reason to be gratified. But how will he reach that happy state? And what will he do when he reaches it— rest on his laurels or continue to write?

These final notes on article writing are intended to help the worker write— and continue to write. Some specific suggestions to these ends are that he:

1. *Continually search for possible topics.* Many employees who write articles get at least one idea a day for a possible subject. On the job, a particular work experience, an observation of another's activity, a conversation with a fellow employee—any one of these may hold one or more topics.

2. *Explore the possibility of writing several articles on the same basic subject.* The author of "Silicones: A Key Ingredient in Aerosol Furniture Polishes," R. J. Thimineur, has written several articles on the subject. One of these, "Silicones in Cosmetics," resulted after publication, Thimineur reports, in "over 150 requests for reprints from approximately 30 countries, including several iron curtain countries." If the aspiring author is employed on projects in which a single material or ingredient is being used again and again, he may well write a series of articles, each on a different application, as Thimineur has done.

3. *Always consider many points of view.* No matter how many articles you see in print that were written by others about *your* subject, try to find a new point of view—a fresh approach—for writing an article of your own. For example, within a few months several articles appeared on the Westinghouse experimental transit system. However, only the article on pages 253–57 was written from the first-person ("I") point of view. Of course, in particular cases many other points of view may come to mind.

As far as the writing experience itself is concerned, suggestions for the worker to make his task easier are that he:

4. *Write regularly.* Writing every day is the best recommendation for getting an article finished—and another started soon after. Also, it is desirable to write at the same time every day. To be sure, temperaments and personal responsibilities vary greatly, and some workers cannot "turn it on" and "turn it off" by the calendar or the clock. But if the individual writes only occasionally, he may soon lose his incentive to finish; he may even abandon the article.

5. *Write rapidly.* Writing fast enables the worker to finish quickly and also to lend a note of immediacy to the article. Furthermore, a point that overlaps item 4 is that he will more likely ensure consistency in point of view, tone, and style than if he writes slowly and infrequently.

6. *Complete his first draft before he attempts to polish the article.* Writing to the end before revising ensures that the article will be completed. Far too many beginning writers fret so much about organizing and expressing their ideas "perfectly" that they become frustrated and quit.

7. *Rewrite.* Rarely does a first effort produce an acceptable manuscript; almost everyone has to rework his early draft, and usually a number of times. Once the first draft is done, the worker should make "writing is rewriting" his motto.

These suggestions will not guarantee either a good article or possible publication. However, they have helped many to be better article writers and to aspire realistically for publication and recognition.

CLEARING ARTICLES WITH EMPLOYERS

The worker should always be sure to obtain his supervisor's permission before he submits for publication anything that is connected with his work. This note is not included to discourage the worker; almost all employers are glad that workers try to write. Indeed, employers realize that publication reflects personal growth, brings credit to the company, and indirectly helps to foster good public relations. But they also worry that confidential information will leak out.

EXERCISES

RECOGNIZING THE BENEFITS OF AUTHORSHIP

1. As you think about the benefits of authorship discussed on pages 247-48, do you consider each to be equally important at the present time? Why? Be prepared to discuss your answers to these questions.

FINDING A SUBJECT

2. Drawing on your experiences (in the laboratory, in class discussion, in a summer job, etc.), list five or more subjects that you believe you are now

qualified—or, after research, would be qualified—to write articles about. Compose titles for these articles.

IDENTIFYING READER GROUPS

3. Using the periodicals indexes in your school library, prepare a list of at least five publications for each of the reader-group classifications identified on page 249. For the reader of technical journals, try to list only those journals that are in your major field.

ADAPTING MATERIAL FOR READERS

4. Survey several issues of at least two journals in your major field and read a total of two articles (your choice) in these issues. Then be prepared to evaluate in class the titles of these two articles.
5. Repeat Exercise 4 but copy and bring to class the introduction and the ending of one article. In class, read the title, the introduction, and the ending aloud and then tell the class how effective or ineffective you think these three elements are—and why.
6. Repeat Exercise 5 but evaluate the author's use of the devices and techniques (such as those listed on page 251 that the author used to develop the body of his article.

SPECIAL PROBLEMS

7. In your school library, look in one of the indexes covering subjects in your field. Then locate two articles on the same subject that were published in different professional journals and read both articles. Finally, write a report comparing the coverage and the development of the articles. Be specific in discussing similarities and differences.
8. For a magazine (such as *Electronic Design*) with which you are very familiar, prepare a publisher's editorial statement similar to that on page 250. Make the statement as complete as possible.
9. Either write a new article or revise a technical report you have recently written, for possible publication in a magazine such as *Electronic Design*. Above the title of the finished article, write the name of the particular magazine you have in mind.
10. Either write a new article or revise a technical report you have recently written, for possible publication in a weekly news magazine or a Sunday newspaper supplement. Above the title of the finished article, write the name of the particular publication you have in mind.

Chapter Twelve

THE JOB APPLICATION

When they are hiring, employers seem to make it easy to get a job. They mail letters and brochures to the placement office at your school. Many advertise in newspapers and in publications like *College Placement Annual*. Quite a few hire agencies to recruit for them year-round, and some even use local radio. In addition, most large firms regularly send interviewers to the campus.

Employers seek you in less direct ways, too. They look you over in summer jobs and in work-study or internship programs arranged with your school. When they are located nearby, they may come to school to give lectures and hold seminars. As they talk, they naturally take advantage of opportunities to tell you about their work and future plans. They may also arrange group tours of their plant, and ask several of you to return alone.

Why, then, study the writing of letters to get a job? When there are so many other ways to contact employers, why apply by mail? After interviews, which are often arranged by phone or wire, many employers use the same means to offer a job. And you, in turn, may phone or wire to accept. Writing letters to employers would seem out-of-date.

But what if the company you want to work for doesn't seek you out? Suppose it doesn't come to your school to look for employees. Or perhaps you don't want a job with any of the companies that send interviewers to your campus. Or it's just possible that none of those that do, get in touch with you after interviewing you at school. Whatever the reason, you may find that *you* have to take the initiative, and that you must write a letter so employers will know about you. These are the assumptions on which this chapter is based.

The first point you should note about applying for a job by mail is that, conventionally, the letter itself is only part of the application. The other part is called a résumé. The résumé is a summary of information about you. You write the letter to introduce yourself (and the résumé), to apply, to highlight your qualifications, and to ask for an interview. You write the résumé to present details of your education, interests, activities, and work experience, along with personal data.

Because the letter introduces the résumé and therefore should not be written until the résumé is finished, first we shall look closely at the latter.

THE RÉSUMÉ

As the following example of a résumé illustrates, this part of the job application is quite objective. (The letter that introduces this example appears on pages 269-70.)

```
PETER WILLIAM HARRIS          RÉSUMÉ                    MAY 1, 1971
University Address                     Home Address (After June 5)
11-B Johnson Hall                      116 North Chilcott Lane
Eastern University                     Allard, New York 13821
College Park, New York 14090           Telephone: 607-399-5428
Telephone: 716-241-8093
```

PERSONAL GOALS

      Primary goal is to be a member of the City Planning Department. Ultimate goal is to be director of city planning.

EDUCATION

      Allard-Kimberly High School, Allard, New York. Graduated June 1964.
      Eastern University, Kimberly Center, Kimberly, New York. Associate Degree in Design and Drafting, June 1969.
      Eastern University, College Park, New York. Expect to receive B.S. in Civil Engineering, June 5, 1971.

      Important courses (at both Kimberly and College Park)

Calculus	Highways and Roads	Soil Mechanics
Chemistry	Logic	Statistics
Civil Engineering	Mechanical Drawing	Structures
Economics	Mechanics	Surveying
Engineering	Physics	Technical Writing
Projects	Psychology	Thermodynamics
Fluid Mechanics	Sanitary Engineering	Traffic Safety

ACTIVITIES AND ORGANIZATIONS

      Student Government Association; secretary, senior year
      Dormitory Government; vice president, senior year
      Chi Epsilon; president, senior year
      Intramural sports

PLANS FOR GRADUATE WORK

      Intend to study for master's degree in the urban-planning field at a nearby college or university offering a night-school program.

WORK EXPERIENCE

Full-time

| Summer 1963 – Fall 1966 | Carpenter apprentice for E. J. Gould, Inc., Allard, New York. Rough and finish carpentry on commercial and residential buildings. Earned journeyman's papers, spring 1966. |

| Fall 1966 – Fall 1967 | Engineering aide, State Department of Highways, Kimberly, New York. Checked plans, kept records, helped in engineering projects, and did general office work. |

Part-time

| Summers of 1968, 1969, and 1970 | Foreman for Allard Construction Company, Allard, New York. Supervised crews representing a variety of building trades. Also in charge of purchases and accounting records. |

PERSONAL DATA

Age, 26; height 6'1"; weight, 185 lbs.; health, excellent.
Unmarried.
Draft exempt.
Hobbies and interests, sports, bridge and pinochle, chess; reading, listening to music, building models.

REFERENCES (BY PERMISSION)

Dr. R. W. Hunt, Professor of Civil Engineering (adviser)
200 Nichols Hall
Eastern University
College Park, New York 14090

Mr. John W. Allard, President (employer)
Allard Construction Company
821 Main Street
Allard, New York 13821

Mr. Ralph T. Northrup (personal)
27 Kingsbury Circle
Allard, New York 13821

This résumé is appealing because it has several "plus" qualities. First, headings have been used; the employer can quickly find what he wants to know. Second, by showing both of his addresses and telephone numbers, Harris has made it easy for the employer to contact him. Third, in addition to reporting the major details of his background, Harris has included personal data and names of references. This information will enable the employer to picture Harris better as a person—as an individual. Fourth, Harris has listed the courses that a department of city planning will want to see. And finally, he has noted not only his goals but also his plans for further study and training.

A good résumé, then, has these elements:

1. A heading
2. A statement of personal goals
3. Information about schooling
4. A list of activities and organizations
5. A statement of plans for additional professional training
6. Details of work experience
7. Personal data
8. Names of references

To help you prepare your own résumé, let's look more closely at each of these elements:

A Heading

The heading is made up of a title (such as "Résumé," "Data Sheet," or "Qualifications for Employment") and your name, address, zip code, and telephone number. If your graduation is near, you will do well to include your home address, as is shown in the example on page 261. Also desirable is the date the résumé is prepared. Then the employer knows how up-to-date the information is.

In presenting your heading, be sure that all lines are balanced on the sheet, well spaced, and neatly typed. A carefully planned heading makes a good "first" impression.

A Statement of Personal Goals

A logical entry just below the heading is a statement of present and future objectives. You show where you want to begin your career and where you hope to be after a few years. Employers like to see that you are realistic about your present ability and also that you have initiative and ambition.

Peter Harris's statement of goals is a good example. Here are two other specimens:

```
    First objective is to serve as a junior chemist in your
Greenville laboratory.  Ultimate objective is to direct one
or more of your research groups.

    Primary goal is to work in your Instrument department.
Eventual goal is to have a responsible position in the
Quality Control section.
```

Information About Schooling

Under "EDUCATION" you logically note first the name and address of each school, the certificate or degree earned, and the year you finished. Then you present the details. Perhaps you will include "Important Courses," as Harris did. It may seem advisable to go further and put down individual course titles. Possibly you will even think it a good idea to separate the courses in your major from those you identify under the subheading "Other Valuable Courses" or "Related Courses."

Naturally, you will focus on giving the employer the information *he* wants. However, you will surely try to make a good impression in giving it. For example, if you've earned high grades in courses in your major, you may decide to show the grades next to the course titles. Or, if these grades are only average but you've done well in other courses, perhaps you can show a fairly high standing in your class. Such details are still factual and therefore objective.

A List of Activities and Organizations

Employers like to see how well rounded you are. You don't have to be the most popular student on campus, but you should show that you can get along well with others. Employers want people who will "fit in" with those who are already working for them. Your participation in activities and membership in well-known clubs, groups, etc., help to define you as a desirable person. You should also note committee work and offices you have held. And you should mention acknowledgments of your fine work, such as being elected a member of an honorary society.

A Statement of Plans for Further Professional Training

A very positive note, to most employers, is the indication that you plan to add to your present knowledge. If you are sincerely interested in improving yourself after you finish school, mention your plans in your résumé. If possible, be specific. Tell the prospective employer what particular course or training program you intend to enroll in. If you are aware of its possible benefits, tell him also how you expect your new knowledge or skill will help both you and him.

Details of Work Experience

Employers want to know about your work experience, whether or not it is in your field. Just the fact that you have worked impresses them. Employers may suspect that you have benefited from these experiences—learning from others as you worked with them, understanding how important it is to be cooperative.

If you *have* worked in your field, that's fine. You have an asset some employers regard highly, and therefore should tell them about it. But even if you haven't had professional experience, you should put more under "WORK EXPERIENCE" than the job name, the dates of employment, and the name and title of your supervisor. If you had responsibilities in your summer job, received raises and promotions, worked for the same employer more than once—these and other facts should be reported.

Personal Data

One thing the résumé to this point has not given the employer is a meaningful description of you. A convention is the listing of personal and physical details— your age, height, and weight, along with an honest statement about your health. Another convention is telling the employer your marital status and the number of any dependents you may have. And male applicants should indicate their draft status or note that they have already been in the service.

Finally, the employer is interested in learning about your leisure-time activities and hobbies. These details, however minor, help him to form a "picture" of you. Since photos can prejudice employers (even though they may think they are unbiased), and therefore may not be included, these few details make up an acceptable substitute.

Names of References

There are also conventions in listing the names of those people who are willing to be references. One convention is that you show the name, title, and address of a teacher in your field, an employer, and a personal acquaintance. The other convention is that you list three references. In your own case, however, you may wish to change both the nature and the number. Under some circumstances, it may even seem best to write "References on Request."

NOTES ABOUT RÉSUMÉ STYLE

Résumé style, you may have noticed, is not complete-sentence style. That is, subjects are left out. The natural use of "I" to start a sentence when one is writing about himself can be distracting when the structure is repeated again and again. If you omit "I," you will seem to be more objective and less self-centered.

Parts of verbs can be left out, too. Peter Harris's résumé illustrates: instead of saying, "I was employed as an engineering aide," he wrote simply, "Engineering aide." Résumé style is concise as well as "objective."

As you know, such omissions are unacceptable in other kinds of writing. Thus you may prefer complete sentences. To do so without appearing to be self-centered, you need only to vary your structures. For example, Harris could have started with a phrase ("In this job, I worked as an engineering aide"). Or he could have said, "Working as an engineering aide, I checked plans, kept records, ..." Of course, other structures can be used.

You can also compromise. That is, you can begin with a sentence in résumé style and then use complete sentences. Harris could have used this technique without making a major change: "Engineering aide, State Department of Highways, Kimberly, New York. In this job I checked plans, kept records, ..."

Whatever style, or combination, you like, try to be consistent. If you are inconsistent, readers may be confused.

USE OF THE RÉSUMÉ INSTEAD OF A COMPANY FORM

Preparing a résumé will help you to fill out application forms. Most companies have printed forms that they ask you to complete so that needed data will be in their files. Very likely you will have to fill out a form even after sending a résumé. Completing a form will be easy if you have a copy of the résumé in front of you. All you do is transfer details to the appropriate spaces on the form.

If nearly all companies have a form, you may wonder why you should write a résumé. One reason for doing so is that few forms have room for all the information you would like to include in a certain space. Another reason is that little provision is made for meaningful explanation and comments. There may not even be provision for some of the information you would like to put down. If you use a form only, then, you may not be able to report details and comments that will help to distinguish you from other applicants.

THE COVERING LETTER

By itself, the résumé means little. What is needed is a letter to introduce it to the employer. Such a message is called a *covering letter*.

Although it serves primarily to introduce the résumé, you should try to make the covering letter do much more. Specifically, you should also use it to accomplish each of the following:

1. Show more than self-interest.
2. Use a "crutch" (a contact) if you can.
3. Apply for a specific job, if it is appropriate.
4. Be confident without being presumptuous.
5. Highlight your qualifications.
6. Request an interview and motivate a favorable reply.

To be original, you should not slavishly follow the order indicated. In accomplishing these essentials, you should plan your letter to suit your individual background, the nature of the job and your qualifications for it, and your immediate and long-range goals. However, there are points of logic to consider. Hence we will discuss the essentials according to where—beginning, middle, or end of the letter—they logically appear.

Make an Attention-Getting Start

Show More Than Self-Interest

If you want very much to work for a particular company or kind of company, why not show your desire in the beginning? Too many applicants seem to be interested only in themselves. They ignore the company's point of view. They fail to realize that the employer wants to know why they are interested in working for him. He may spend a small fortune on training new employees, only to find that employees seek another job when the training program is over. You can understand, therefore, why the employer is impressed by applicants who are interested in his company for its own sake.

Of course, any expression of interest should be sincere—as this beginning illustrates:

```
    Last summer I had the privilege of working in your
Process Design Department.  I was very interested in the work
that I was doing, especially the problem of designing the
new heat exchanger that Mr. Warren was attempting to solve.
I should like very much to work permanently in a department
like his, and am therefore applying for the position of
assistant engineer that I understand is open there.
```

If you cannot honestly identify with the company, surely you can express your interest in the job, your profession, or the industry. You should say what *you* mean—not perfunctorily copy the objective writing of an applicant who doesn't care where he works, as long as he gets a job.

Use a "Crutch" If You Can

A *crutch opening* is one that takes advantage of a contact you can make with a particular company. The crutch may be a reference to

1. A person whom the reader knows or will respect.
2. A published notice of an opening.
3. A job experience with the company.
4. A plant tour of the company.
5. A speech or article about the company.
6. An experience like growing up near the company.

The beginning just shown illustrates how the third reference might be developed. Here are examples of the first and fifth:

```
    Dr. Harold Lewis, your Senior Research Director, informed
me today that there are openings in the job of chemical
technician at your Biological Research Laboratory in
Bethesda.  I should like to apply for one of the positions
because of my interest and experience in both organic and
inorganic laboratory work.
```

```
    At the December meeting of the American Institute of
Chemical Engineers, I had the opportunity to hear Dr. Ralph
E. Wilson speak on your exploratory research programs in
surface chemistry and catalytic reactions.  His talk, to-
gether with the knowledge of catalysis I have gained through
course study, leads me to apply for a position as research
trainee.
```

Apply for a Specific Job

A good opening paragraph shows interest in a particular job or kind of job, as the preceding examples show. Then the writer can move logically into the paragraphs that present his qualifications.

Some applicants like to begin with a single-sentence paragraph that gets directly to the point:

```
I should like to apply for the job of process engineer
at your Maryville plant.
```

Although this approach gets the applicant off to a running start, you may think it is too abrupt. If you prefer a longer introduction, then write one. Or use a beginning that is direct, and then elaborate in the same paragraph. But no matter how you start, always try to indicate early in your letter the job or kind of job you want. The employer won't know if you don't tell him.

Be Confident Without Being Presumptuous

One way the employer can determine how you will fit into his company is by the tone you use. Naturally, one judges you by more than the tone of your letter. However, the tone may influence the reader greatly, since it reveals the way you look at yourself, at your career, at employment in general, and—especially—at him. The tone of your writing indicates your attitude toward what you are writing about.

What tone will satisfy the reader? You need to show, by your ideas and how you present them, that you are mature, realistic, sincere, and considerate. You also need to show that you are ambitious and self-confident. Yet you must not seem to be self-righteous; nor, conversely, should you seem to be self-effacing.

Give the employer reason to believe what is probably true—that you are "somewhere in between." Showing that you are confident is a *must,* but just a little modesty will go a long way. Prefer using phrases like "I am confident," "I think," and "I believe," to "I know," at the one extreme, or "I hope I can live up to . . . ," at the other. "I know" and similar phrasing make your ideas sound categorical; "I hope . . . ," too self-effacing. What might be called "the right tone" can be found in the specimen paragraphs to this point in this chapter.

Highlight Your Qualifications in the Middle Paragraphs

The middle paragraphs are the heart of your letter. Here you support your assertion that you "believe" you are "qualified for the position that is open." You should not go into great detail, because, after all, the résumé presents the details. But you need to convince the employer that you have the ability to do the job as he wants it done. Indeed, displaying some humility, you should try to convince him that you can do the job better than he expects.

Most applicants point up their qualifications as they think best. They should not develop their letters according to any "formula" plan of development; rather they should follow the "direction" indicated by their résumé. Of course, the things you should talk about are your education, work experience, activities, and interests. But in the heart of your letter, as in the beginning and the end, you need to keep in mind the employer's point of view. How you develop your letter depends on your analysis of the individual situation.

One paragraph, the second in a letter by a student aspiring to be a plant pathologist, should illustrate:

Last summer I worked as a research assistant under Dr.
John S. Warren, a virologist in the Plant Pathology Department here at Midland University. During that time I worked
not only with viral diseases but also with various fungal and
bacterial diseases, especially those peculiar to tomatoes and
corn. One of the projects, which your company sponsored, involved testing tomatoes for the cause of internal browning, a
disease which you know has plagued the industry throughout
the East for several years.

Request an Interview and Motivate a Reply at the End

Conventionally, in the last paragraph the applicant asks for an interview and tries to motivate the employer to grant one. Logically, therefore, the paragraph will be short. However, you should not make it so short that the reader merely glances at it. You should seek to hold his interest to the end. This closing paragraph is a good example:

Your laboratories are well known for their continuing
achievements in medical research. I am sure that these
successes would have been realized with difficulty if the
atmosphere were not favorable. I would like very much to
work in such an atmosphere, and would appreciate having an
interview at your convenience. Since my home is nearby,
I will be able to come to Millikan any Saturday you wish.

To see how all these essentials can be included effectively, let's look at the letter Peter Harris wrote to introduce his résumé. Note the ways in which he attempted to be original:

11-B Johnson Hall
Eastern University
College Park, New York 14090
May 1, 1970

Mr. George E. Latham, City Manager
City Hall
Main and Elm Streets
Macon, Pennsylvania 19801

Dear Mr. Latham:

During our field trip to Macon last fall, I was greatly
impressed by the city's growth potential and the many opportunities available to young people there. Having a strong
interest in urban planning in such a community, I would like
to apply for the position of junior engineer in your City
Planning Department. In suggesting that I write, my advisor,

Professor R. W. Hunt, told me that this new job is to be filled by the end of this summer.

Professor Hunt has helped me to select course work that should be valuable in urban planning. This year, for example, I have had courses in general construction methods and cost estimating, highway safety and traffic control, and street and highway design. In these advanced electives in Civil Engineering, I have acquired a great deal of up-to-date information for use in contributing to solutions of problems much like those Professor Hunt says Macon is now facing.

To supplement these and other courses, I have earlier educational benefits and a variety of experiences that you may consider desirable. Before coming to College Park, I completed the University's two-year program in Design and Drafting at Kimberly, where I had enrolled with the goal of being a major contractor. Prior to this, I had worked for several years--first as a journeyman carpenter and later as an engineering aide in the State Department of Highways. And for the last few summers I have been job engineer and foreman for a large contracting firm in my home town. Through all of these experiences I learned much about the viewpoints of those who work regularly with planners of federal, state, and city projects.

At Eastern I have been in as many organizations and activities as possible, including the Student Government Association, my dormitory's governing unit, the Civil Engineering honorary, and intramural sports. In several of these I have held positions of leadership.

Mr. Latham, I have tried here to introduce myself as I believe you would like to know me. Perhaps the details in the résumé enclosed will answer any questions you have about my qualifications and my interest in helping to solve Macon's urban-planning problems. However, I would like very much to discuss with you this job that has been newly created in your city, and would appreciate having an interview at your convenience.

Sincerely yours,

Peter W. Harris

Peter W. Harris

This letter and the résumé on pages 261-62 should aid you in answering any questions you have on job applications. Neither letter nor résumé should be merely a patching together of parts of textbook samples. Employers like to see you follow conventions, to be sure, but they also like to see evidence of imagination and originality.

FOLLOW-UPS

The process does not end with the mailing of your letter and résumé. In fact, it has just begun. You may have to remind the employer that you have applied; you will definitely have to prepare for the interview; and eventually you will have to accept or reject the offer if one comes. Although these are important steps in the process, we shall look at them only briefly. What you have learned thus far should guide you well when you face each of these situations.

Consider writing a follow-up, or reminder, letter if you receive no answer to your application within a reasonable time. This may be as short as two weeks and as long as a month or more; your distance from the employer will help you decide what "reasonable" means. The letter itself should be brief. You want to remind the employer but not to annoy him. What you *need* to say, of course, is that you are still interested and will appreciate an early response. Concisely reviewing qualifications and your desire for an interview is also expected and in good taste. Then stop—without being abrupt, naturally. There is no reason to send another copy of the résumé.

In the interview, try to be as courteous and sincere as you attempted to be in your application. Be alert, and as candid as you can be without being negative about what you have reason to believe the employer expects to be treated in a positive way. Common sense should tell you not to say anything just because you *want* to say it: say what is on your mind only because there is *reason* to say it. In other words, mind your thoughts as well as your manners. As far as your appearance is concerned, surely you will look as well groomed as you can.

Whether or not you write a letter after the interview will likely be determined at the interview. If the employer indicates that you should not write, but should instead wait for him to contact you, then you will not write. Otherwise you should. It's always in good taste to write a "thank you" note, especially if a good deal of money and time has been spent on you. Furthermore, the visit to the company may well have intensified your interest in working there. It would be appropriate—and probably gratifying to the employer—for you to say so.

Accepting a job offer is easy; it's always easy to write a letter that you very much want to write. You should do more than accept and express your thanks, though. For example, there are usually papers to send, or you need to confirm the starting date or some other arrangement, or you'd appreciate help in finding temporary housing. And you may want to make a comment or two to show your enthusiasm in looking ahead to the job experience.

Refusing an offer is not so easy. It's easy enough to begin with a statement of thanks for receiving the offer; but then finding a way to say "no" graciously poses a problem. After all, you want the employer to think well of you—and you *don't* want him to think his company is "second best" in your judgment. To say "no" effectively (and comfortably), explain why you are declining the offer before you actually refuse it. And be careful how you say "no": think twice, for example, before saying, "I must decline your offer." The word "must" can hardly be called logical if your reason for refusing is simply a preference for another company's offer.

Finally, remember that it is good manners to let other companies know your decision, even when you haven't yet received offers from them. A few words of appreciation for considering you, along with a note that you have accepted an offer, are always gratifying to other employers. The information helps them greatly in their own recruiting work.

EXERCISES

THE RÉSUMÉ

1. Discuss the effectiveness or ineffectiveness of the following excerpts from résumés written by students applying for their first permanent job.

   a. From the résumé of a student applying for a job as a guide for the National Park Service:

   ```
   Have had a variety of summer jobs ranging from
   selling soft ice cream to teaching conservation educa-
   tion for a state's department of conservation.
   ```

   b. From the résumé of a student applying for a job as a petroleum engineer with a large oil company:

   ```
   Have had some experience in the field, working as
   an engineering trainee with your company during the
   past three summers.
   ```

   c. From the résumé of a student applying for a job as a junior engineer in the research and development section of a small engineering firm:

   ```
   My first objective is to work as a junior engineer
   in your research and development department.  My ul-
   timate objective is to be a manager in a company that
   specializes in research and development.
   ```

THE COVERING LETTER

2. Discuss the effectiveness or ineffectiveness of the following excerpts from letters of students applying for permanent jobs.

   a. The first two paragraphs of a letter addressed to the director of personnel at a large oil company:

   ```
   While I was in the University Placement Office last
   week for the first time, I learned of the possibility
   of openings in your research and development division
   for chemical engineers and chemists.  I am intensely
   ```

interested in this branch of industry and would like
very much to work for your company.

  I will be graduated next week from Central University with a B.S. in Chemical Engineering. At present
I am in the upper half of the third fifth of my class.

b. The first two-plus paragraphs of a letter addressed to "Chief, Branch
of Personnel, Bureau of Indian Affairs":

  An article in this year's College Placement Annual
points out that your bureau is concerned with land
use and the management of natural resources. Emphasis
is placed on those services which are most beneficial
to the American Indians. I, too, am concerned. Today's
problems in forest land management are of great interest
to me, especially in the field of watershed management.

  I will graduate from Southwestern University in
June with a B.S. degree in Forest Technology. This
major provides specialized training important to the
proper management of forest lands.

  During the past two years I have gained experience
and insight in the field of watershed management by
working as a research aide under the direction of Dr.
Vernon J. Miller. . . .

c. The first paragraph of a letter addressed to the director of the Data
Automation Division, National Meteorological Center:

  After being employed last summer in your division,
I feel that I have developed an interest in the computer work connected with meteorology. Throughout
this year I have taken several more courses in
meteorology which gave me a broader scope of the field.
With this additional background I now feel better
qualified to handle similar work in a permanent job
with the Data Automation Division.

3. Discuss the effectiveness or ineffectiveness of the following paragraph—
the second in a three-paragraph letter written by a student who is applying for admission to a school of veterinary medicine:

  I will complete my third and final year of pre-
veterinary medicine at Compton University in June,
along with thirty other students. At present I am
twentieth in my class with a 3.00 all-university
average. Being an active member of Compton's large
Pre-Vet Club, I have had the opportunity to meet and
talk with several prominent veterinarians about
veterinary schools and the overall field of veterinary
medicine. It has also given me the chance to tour the
facilities of the University of Pennsylvania and those

of your school, and to meet and speak with some of the
personnel. Along with being a member, I also serve as
one of four representatives for the Pre-Vet Club on
the Agriculture Student Council. My interest in dairy
cattle has moved me to join and maintain two years
active membership in the Compton University Dairy
Science Club. I am also a brother of Gamma Zeta, the
agriculture fraternity at Compton, and a member of
the Bridle Club. I participate regularly in intramural
football, basketball, bowling, soccer, and volley ball.
I also have a vivid interest in hunting, fishing,
hockey, and trapping.

4. Discuss the effectiveness or ineffectiveness of the following complete job
    application addressed to the personnel director of a large company special-
    izing in computer services to industry and science.

Dear Mr. Purdy:

   My degree in meteorology and twelve credits in computer science seem to make me a good prospect for your
job opening dealing with the use of computers in science
and industry.
   I will graduate this spring from Southeastern University with a B.S. degree in meteorology. During my
studies here I became interested in computer science
and took four three-credit courses on the subject.
   If possible, I would like to have an interview with
you at your convenience.

                              Sincerely yours,

                              Frank W. Smith

Enclosure: Data Sheet

                  * * * * * * * * * *

                  Frank W. Smith - Data Sheet
                     7711 West Cornwall Avenue
                     Carruthers, Georgia 30371

Personal Details

     Age 22, unmarried, draft exempt, height: 6 ft 1 in.,
       Weight: 173 lb
     Interests: judo, weight training, basketball,
       photography

Education

> Carruthers Senior High School; graduated 1967
> Southeastern University; B.S. in meteorology,
>   June 1971
> Grade-point average: 2.46
>
>> Important courses (and credits)
>> Mathematics (31)   Computer Science (12)
>> Meteorology (21)   English (9)
>> Physics (16)   Psychology (6)
>> Chemistry (6)

Activities and Organizations

> Student Council member, College of Science
> Member, Southeastern Judo Club

Work Experience

> Summer 1967: refuse collector
>   Employer: City of Carruthers
> Summer 1968: construction worker
>   Employer: Mr. John Luther, Carruthers
> Summer 1969: laborer
>   Employer: Carruthers Metal Works, Inc.,
>   Carruthers
> Summer 1970: mail carrier
>   Employer: Post Office, Carruthers
> Christmas employment, 1968–70: mail clerk
>   Employer: Post Office, Carruthers

References (Supplied on Request)

5. Discuss the effectiveness or ineffectiveness of the follow-up letter which follows, written by Frank W. Smith two weeks after he mailed his résumé and covering letter to the company identified in Exercise 5.

Dear Mr. Purdy:

> Two weeks ago I sent my résumé and letter applying for a job with your company.
> I am now even more interested in a position with your company and hope that my application is still being considered.
> Please let me know your decision concerning me as soon as possible.

SPECIAL PROBLEMS

6. Write a résumé and a covering letter for an application for permanent employment. Assume that the time is just before your graduation.

7. Write a résumé and a covering letter for summer employment in your field of study.
8. Assume that two weeks have passed since you wrote the job application assigned in Exercise 8. Now write a follow-up letter.
9. Assume that you have received a job offer. Write a letter of acceptance.
10. Assume that you have received a job offer. Write a letter of refusal.

A TECHNICAL WRITER'S
HANDBOOK

Before submitting a report, memo, letter, etc., you should spend time reviewing your writing for weaknesses pointed out in the first four parts of this book. In addition, you should appraise the basic English in your writing. To help you edit for errors, this handbook is divided into three sections. First covered are conventions such as abbreviations, numbers, and spelling. Then you will find explanations of terms like *noun* and *verb*, to aid you in your use of the third section, which covers common errors in diction, grammar, and punctuation.

To aid you in finding individual errors in your writing, faults are classified (for example under "Spelling") and numbered within the classification. Abbreviations (for example, "SP" for spelling) are used, as well. The complete classification, along with page numbers to show you where each section begins, is as follows:

(AB)	Abbreviations, p. 280
(NUM)	Numbers, , p. 287
(SP)	Spelling, p. 288
(SU)	Violations of Sentence Unity, p. 303
(SC)	Violations of Sentence Coherence, p. 305
(D)	Errors in Diction, p. 313
(P)	Errors in Punctuation, p. 315

To aid you in understanding most of the grammatical terms used in the book, a list of explanations is included. Its listing is shown as follows:

(GT)	Explanations of Grammatical Terms, p. 293

Also, you will note that in the sections "Spelling," "Violations of Sentence Unity," "Violations of Sentence Coherence," "Errors in Diction," and "Errors in Punctuation," questions appear in subclassifications. Questions (such as "Is the Sentence a Fragment?") are used because they are more easily remembered than topic statements (such as sentence fragment) and therefore should help you to look for individual errors. An abbreviation (like *frag* for sentence fragment) or a symbol appears after each subclassification, too.

To extend this system to areas of organization and style, you may wish to note the following references to pages earlier in the text that have to do, in general, with writing itself. "O" stands for "Organization," and "S" stands for "Style":

(O 1) Are order-of-development forms used logically and effectively? (dev) (See p. 29.)

(O 2) Is the outline (Are headings) logical? (plan) (See p. 31.)

(O 3) Is paragraphing logical and effective? (¶) (See p. 34.)

(S 1) Is the sentence (Are the sentences) effective? (sent) (See p. 44.)

(S 2) Is the sentence mood appropriate? (sm) (See p. 49.)

(S 3) Is the sentence voice effective? (sv) (See p. 50.)

(S 4) Is the sentence order effective? (so) (See p. 52.)

(S 5) Is the handling of phrases effective? (phr) (See p. 53 and p. 299.)

(S 6) Is awkwardness apparent? (k) (See p. 54.)
(S 7) Is the word choice appropriate? (wc) (See p. 55.)

ABBREVIATIONS (AB)

The following conventions are observed in most technical, scientific, and business writing. After the list of *do's* and *don't's* you will find a list of common technical abbreviations.

When to Abbreviate

1. Abbreviate the titles *Mr., Messrs.* (plural), *Mrs., Mmes.* (plural), *Ms.* (when you don't know a woman's marital status), and *Dr.*
   Examples: *Mr. Frank Klemans, Ms. Jane Knox, Dr. Paul Lowenstein.*
2. Abbreviate *Junior* and degrees such as *Doctor of Philosophy* and *Doctor of Medicine* when they follow a proper name.
   Example: George W. Donaldson, *Ph.D.*
3. Abbreviate the names of organizations and agencies for which the abbreviation is commonly used.
   Examples: *ALCOA* for Aluminum Company of America, *TVA* for Tennessee Valley Authority. (Note that no periods appear after the letters.)
4. Abbreviate in footnotes, bibliography, references, etc. (See pp. 79–82. Note, however, that only certain words may be abbreviated.)
5. Abbreviate the directions of the compass and of lines of longitude and latitude.
   Examples: *NE, SW, 48° N, 13° W*
6. Abbreviate the words for mathematical relationships when writing equations and formulas, but always explain for nonspecialists before or after presenting equations. For these readers, the following example should be written as "*A* is greater than *Q.*"
   Example: $A > Q$.

When Not to Abbreviate

1. In important correspondence (letters of inquiry, job applications, etc.) do not abbreviate titles (*Manager,* not *mgr.*), street directions (*North Main Street*, not *N.* or *No. Main St.*), or cities or states (*Los Angeles, California,* not *L.A., Cal.*).[1]
2. Do not abbreviate names of the month (write *September*, not *Sept.*) or days of the week (write *Friday* not *Fri.*).

[1]When you are paying personal bills, writing to friends, etc., the use of abbreviations is both conventional and very much in order. The Post Office Department, experimenting with a scanner which will simplify mail distribution, may, in the near future, ask everyone to abbreviate states in addresses with two capital letters (see the table on p. 154).

3. Do not abbreviate first names (write *Charles*, not *Chas.*) or any word that is never abbreviated in standard English usage (like *calculus*).
4. Do not abbreviate any part of an organization's name (write *Company*, not *Co.*; write *Incorporated,* not *Inc.*). Naturally, if the word is abbreviated as part of the organization's legal name, it should not be spelled out.
5. Do not abbreviate common words like conjunctions (write *and*, not +; write *accordingly*, not *acc'ly*). Except in mathematical equations, formulas, etc., write out therefore (∴), since, because (∵), etc.

Notes on Abbreviations

The following points are important:
1. The hyphen is included in abbreviations for compound attributive adjectives (adjectives before the noun that are hyphenated).
   Examples: *3/4-inch high-pressure copper tubing* in abbreviation becomes *3/4-in. h-p copper tubing; alternating-current circuit* becomes *a-c circuit.*
2. Technical abbreviations have no plural form.
3. Monetary symbols are conventional "abbreviations" ($, ¢), as are the symbols for percent (%), *by* (2 × 4), feet ('), and inches (") in measurements, and minutes (') and seconds (") in time.
4. Periods appear after only those technical abbreviations that can be read as standard words (*at.* for *atomic, bar.* for *barometer, in.* for *inch,* and *long.* for *longitude*).

Common Technical Abbreviations

Although not all-inclusive, the following list of technical abbreviations does cover several fields. If an abbreviation you are looking for is not in the list, in your dictionary, or in a style manual published by a journal in your field, spell out the word(s). Note that only five hyphenated compound words (the adjectives alternating-current, direct-current, high-pressure, and low-pressure, and the proper name Pensky-Martens) appear in the list. If you wish to use other hyphenated compounds (example: gram-calorie) that are common, be sure that the individual words have abbreviations and remember to include the hyphen in the abbreviation (example: g-cal).

absolute	abs
acre	A
alcohol or alcoholic	alc
alkaline	alk
alternating-current (as an adjective)	a-c
American melting point	AMP
American wire gage	AWG
ampere	amp or A
Angstrom unit	Å
are (square dekameter)	a. (dkm^2)

atmosphere	atm
atomic weight	at. wt
average	av
avoirdupois	advp
barometer	bar.
barrel	bbl
biochemical oxygen demand	BOD
Birmingham wire gage	BWG
board foot	bd ft
boiling point	bp
brake horsepower	bhp
Brinell hardness number	BHN
British thermal unit	Btu
Brown & Sharpe (gage)	B&S
bushel	bu
calorie	cal
candela	cd
candlepower	cp
centare (square meter)	ca (m^2)
center to center	c to c
centigram	cg
centiliter	cl
centimeter	cm
centistoke	cs
chemical	chem
chemically pure	cp
chemical oxygen demand	COD
concentrate	conc
conductivity	cond
cord	cd
coulomb	C
counter electromotive force	cemf
cubic centimeter	cc or cm^3
cubic decimeter	dm^3
cubic dekameter	dkm^3
cubic feet per minute	cfm
cubic feet per second	cfs
cubic foot	cu ft
cubic hectometer	hm^3
cubic inch	cu in.
cubic kilometer	km^3
cubic meter (stere)	m^3 (s)
cubic micron	μ^3

cubic millimeter	mm^3
cubic yard	cu yd
cycles per minute	cpm
cycles per second (Hertz)	cps (Hz)
cylinder	cyl
decibel	dB
decigram	dg
deciliter	dl
decimeter	dm
degree	deg or °
degree Baumé	°B
degree Celsius (centigrade)	°C
degree Fahrenheit	°F
degree Kelvin	°K
dekagram	dkg
dekaliter	dkl
dekameter	dkm
diameter	diam
direct-current (as an adjective)	d-c
dram	dr
effective horsepower	ehp
efficiency	eff
electric	elec
electromotive force	emf
electron volt	eV
elevation	el
European melting point	EMP
evaporate or evaporated	evap
extreme pressure	EP
farad	F
feet per minute	fpm
feet per second	fps
fluid	fl
foot	ft
freezing point	fp
fusion point	fnp
gallon	gal
gallons per minute	gpm
gallons per second	gps
gauss	G
gram	g or gm
gravity	gr

hectare (square hectometer)	ha (hm^2)
hectogram	hg
hectoliter	hl
hectometer	hm
henry	H
Hertz (cycles per second)	Hz (cps)
high-pressure (as an adjective)	h-p
hogshead	hhd
horsepower	hp
hour	hr or h
hundredweight (100 lb in the U.S.; 112 lb in England)	cwt
hydrogen ion concentration (the negative logarithm of the hydrogen ion concentration of a solution)	pH
inch	in.
inches per second	ips
indicated horsepower	ihp
infrared	IR
inside diameter	ID
insoluble	insol
joule	J
kilocalorie	kcal
kilocycle	kc
kilogram	kg
kilograms per second	kgps
kilohm	k Ω
kiloliter	kl
kilometer	km
kilometers per second	kmps
kilovolt	kV
kilowatt	kW
Lambert	L
latitude	lat
linear foot	lin ft
liquid	liq
liter[2]	l
longitude	long.
low-pressure (as an adjective)	l-p

---

[2]Prefer not abbreviating "liter"—if only because it may be confused with the numeral 1.

lumen	lm
lumens per watt	lpW
maximum	max
mean effective pressure	mep
melting point	mp
meter	m
microampere	μA
microfarad	μF
microgram	μg
microinch	μin.
micromicrofarad	$\mu\mu$F
micromicron	$\mu\mu$
micron	μ
microvolt	μV
microwatt	μW
miles per hour	mph
miles per hour per second	mphps
milliampere	mA
millifarad	mF
milligram	mg
millihenry	mH
milliLambert	mL
milliliter	ml
millimeter	mm
millimicron (nanometer)	mμ (nm)
million electron volts	MeV
million gallons per day	mgd
milliroentgen	mr
millivolt	mV
milliwatt	mW
minimum	min
minute	min
molal (molal solution)	$\underline{m}$
molar (molar solution)	$\underline{M}$
molecular weight	mol wt
nanometer (millimicron)	nm (mμ)
Newton	N
normal (normal solution)	$\underline{N}$
nuclear magnetic resonance	NMR
Number (as in Number 1)	No.
oersted	Oe
ohm	Ω

Open Cup (Flash)	OC
ounce	oz
outside diameter	OD
parts per billion	ppb or PPB
parts per million	ppm or PPM
peck	pk
pennyweight	dwt
Pensky-Martens (Flash)	P-M
pint	pt
poise	P
pound	lb
pounds per square inch	psi
pounds per square inch gage	psig
quart	qt
radian	rad
relative humidity	RH
revolutions per minute	rpm
revolutions per second	rps
roentgen	R
root mean square	rms
Saybolt Universal Viscosity	SUV
second	sec or s
shaft horsepower	shp
soluble or solution	sol
specification(s)[3]	spec(s)
specific gravity	sp gr
specific heat	sp ht
square centimeter	cm^2
square decimeter	dm^2
square dekameter (are)	dkm^2 (a)
square foot	sq ft
square hectometer (hectare)	hm^2 (ha)
square inch	sq in.
square kilometer	km^2
square meter (centare)	m^2 (ca)
standard	std
standard temperature and pressure	STP
stere (cubic meter)	s (m^3)
Tag Closed Cup (Flash)	TCC
Tag Open Cup (Flash)	TOC

---

[3]Generally, this abbreviation is used only in informal intracompany writing.

temperature	temp
tensile strength	tens str
tetraethyllead	TEL
thousand electric volts	keV
thousand gallons per day	tgd
ultraviolet	UV
United States (gage)	US
United States Pharmacopoeia	USP
vapor temperature	vt
viscosity	vis
viscosity index	VI
volt	V
volume percent	v%
watt	W
weber	Wb
weight	wt
weight percent	wt%
yard	yd
year	yr

NUMBERS (NUM)

In your earlier schooling, you probably learned to use Arabic numerals (the figures 1, 2, 3, etc.) only when more than two words were to be used for whole numbers. In technical writing, however, you should use Arabic numerals for all precise numbers over nine.

You should also use figures in the following places, whether or not the number is more than nine:

1. The day of a date (September 28, 1971, your memo of April 9) and numbers used for time (8:30 A.M, at 4 P.M.).
2. The number of an address (2176 East River Road). Any street name that includes a number under nine is always written out (757 Third Avenue but 452 10 Avenue).
3. Measurements (6 in. long for 6 inches long), decimals (0.5 gps for 0.5 gallons per second), amounts of money ($9.28), and percentages (3%).
4. In a series (. . . 3 journeymen, 12 apprentices, and 2 foremen).

However, you should never begin a sentence with figures, even when figures would be used ordinarily:

Wrong: 170° was the temperature recorded.

Right: One hundred seventy degrees was the temperature recorded.

Right (*and preferred*): The temperature recorded was 170°. (Note that the words here are placed in a different order, so that Arabic numerals can be used.)

Notes on the Use of Numbers

1. When a decimal contains no whole number, put a zero in front of the decimal point. Examples: 0.12 g, 0.52%
2. When a table or list contains both whole numbers and decimals, always align on the decimal point (even though no decimal point is used after a whole number):

   Example: Color, ASTM D 1500-58T   7.5

              Water: PPM   320

              Agent 046: %   0.52
3. Use Arabic numerals for fractions and mixed numbers, except when fractions do not modify directly a noun of measurement, amount, etc.
   Examples: $\frac{1}{4}$-ft lengths, $1\frac{1}{2}$-in. pipe
   but: one quarter of a turn, five and one-half inches of water
4. Use Arabic numerals and words for large numbers; use numerals only in tables and figures.
   Examples: 500 million units, $125 million
5. For important numbers (e.g., in contracts, legal documents of other kinds) both write out numbers and use Arabic numerals in parentheses.
   Example: The total price of the installation is to be nine thousand sixty dollars and sixteen cents ($9,060.16).
   Arabic numerals alone would be appropriate here, but if an error in typing were made, the result could be disastrous!

SPELLING (SP)

Because the rules permit many exceptions, this section on spelling is largely a reminder. Also included is a list of words that many writers tend to misspell. Writers misspell for many reasons, of course, but the major reason is carelessness.

Other errors covered are the misuse and the omission of the apostrophe and the hyphen, and the use and the misuse of capital letters. These errors are not always classified as spelling errors; however, they are so labeled here because they are different in form from what they should be. Any error in presenting the form of a word can properly be called misspelling.

In reviewing the following list of words that are commonly misspelled and checking your own spelling, ask yourself these questions:

1. When I pronounce the word, does it sound as it is spelled (for example, *perform*)?

2. Have I thoughtlessly written a homophone? A homophone is a word that sounds like the word wanted, but it is spelled differently and has a different meaning. (*Cite, sight,* and *site* are all homophones.)
3. Have I carelessly omitted, transposed, or added letters (for example, *goverment, thier, reguard*)?
4. Have I misheard and therefore misspelled words that are included in a common expression (for example, *for all intensive purposes* instead of "For all intents and purposes," *per say* instead of "per se," *rightaway* instead of "right of way," and *sort of speak* instead of "so to speak").

Some of the words you may have trouble with can be easily checked if you exaggerate the pronunciation of the questionable vowel. For example, you can exaggerate the pronunciation of affect as *ay*ffect, or effect as *ee*fect. (Of course, you should exaggerate only for the sake of remembering which word has which vowel.) When you visualize or actually *write* the word, you should spell accurately.

Also, some words may be easier to spell if you think of another word that comes from the same root. For example, it is easier to remember that *permissible* has an *i*–rather than an *a*–before the *b* if you think of *mission*. Both words come from the same Latin root. Finally, you may find helpful the use of a gimmick. Take *source*, for instance. If you tell yourself that you always want to "see" the source, you'll never write *sourse*. Or take *sincerely*. If you tell yourself that anybody can *rely* on you, you'll never forget the second *e*.

When you review your reports, letters, etc., just before writing the final draft, a "last-minute-check" technique you may find helpful is reading lines *backwards*. When you read from left to right, you may overlook misspelling because you are naturally conscious of the sense of your writing. However if you examine lines *from right to left*, you will be more conscious of individual words than of what the words, together, are saying.

(SP 1) Are These Words Misspelled Because of Carelessness? (list)

acceptable	advantages	apologize	calendar	concede
accessible	adversely	appropriate	catalog	concession
accessory	advice	argument	(catalogue)	conducive
accidentally	advisable	assurance	category	congratulate
accommodate	affect	attendance	changeable	conscientious
accurate	aggravate	authorize	cite	conscious
accustom	agreeable	a while	column	consistent
achievement	allotted	balance	coming	controlled
activities	all ready	becoming	commission	convenience
addressed	already	before	commitment	courteous
adequate	among	believe	comparable	criticism
adjustment	analysis	beneficial	competitive	criticize
advantageous	analyze	benefited	competitor	curriculum

deductible	government	occurred	quantity	subsidiary
deferred	grateful	occurrence	questionnaire	subsidize
deficiency	grievance	omission	receipt	succeed
definite	guarantee	omitted	receive	success
desirable	helpful	opinion	recommendation	summary
desperate	hesitant	opportunity	recur	surprise
deterrent	immaterial	optimistic	reference	tendency
disappoint	implement	original	referred	than
discrepancy	indispensable	pamphlet	regard	their
discriminate	inducement	parallel	reimburse	then
dissatisfied	innovation	perform	relevant	there
distribution	in order	permanent	remittance	thorough
effect	installation	permissible	repetition	through
eligible	integrate	personal	respectfully	to
embarrass	intercede	personnel	respectively	too
enterprise	laboratory	pertinent	rhythm	transfer
environment	led	plausible	safety	transferred
equipment	liability	possession	salary	truly
equipped	liable	practical	schedule	undoubtedly
equivalent	maintain	precede	seize	unnecessary
eventually	maintenance	predominant	separate	useful
evidently	manufacturer	preference	sight	using
exaggerate	merchandise	preferred	similar	vacuum
exceed	miniature	principal	sincerely	various
familiar	miscellaneous	principle	source	warrant
feasible	necessary	privilege	sponsor	weather
forfeit	negligence	procedure	stationary	whether
formally	negligible	proceed	stationery	wholly
formerly	noticeable	professor	stretch	whose
fulfill	obsolete	prominent	studying	writing

(SP 2) Has the Apostrophe Been Omitted or Misused? (/')

The apostrophe is the mark of punctuation that is easily overlooked by many students. Because the meaning of the word is changed without the mark, however, you should not leave it out. The apostrophe should be used to show ownership except in the possessives *its, his, hers, ours, yours,* and *theirs.* Of course, you should always be careful to position the apostrophe in the logical place: before the *s* for most singular words; after the *s* for most plural words.

> *apostrophe omitted:* The companys experimental work in this field is very well known.
> *correction:* The *company's* experimental work in this field is very well known.

apostrophe misplaced: All employee's records have been brought up-to-date.
correction: All *employees'* records have been brought up-to-date.

You should also be sure that you have not used a plural form of a word—in place of the possessive singular form—to show ownership or possession. This error is common because the plural form *sounds* like the possessive singular.

incorrect use of plural for possessive: The companies experimental work in this field is very well known.
correction: The *company's* experimental work in this field is very well known.

Other uses of the apostrophe are to show that letters are omitted or that Arabic numerals, letters, etc., are to be read as words.

Examples: *It's* [It is] a unique test.

We *won't* [will not] finish our work by the deadline.

I have earned *A's* in all of my major courses.

The decline in production began in the late *1960's.*

(SP 3) Has the Hyphen Been Omitted or Misused? (/–)

In no other writing are compound adjectives—such as *dust-control* in "dust-control equipment"—used more than in technical and business writing. The reason for the use is that industry, science, and business continually create new words by putting "old" ones together. Thereby they are able to describe new developments, combinations of materials and chemicals, processes, theories, etc.

Compounding words can create a problem. Consider, for example, the following sentence:

For this experiment 10 gallon containers are needed.

Does the writer of this sentence mean *10 one-gallon containers* or *more than one container of ten-gallon capacity?* If he means the former, no hyphen should be inserted between "10" and "gallon." However, if he means the latter (more than one container of ten-gallon capacity), the hyphen *must* be inserted between "10" and "gallon."

Note that when the modifying words appear *before* the noun or pronoun, it is never wrong to hyphenate. (Thus, one would do well to write "10 one-gallon containers" if that was what he meant. Indeed, he would clarify "10 gallon containers" in the sentence above if he wrote, "For this experiment 10 one-gallon containers are needed.") When the modifying words *follow* the noun or pronoun, the hyphen is not used:

hyphen needed: filled-label file, six-foot column

hyphen not needed: file of filled labels, a column six feet tall

Exception: All employees' records are *up-to-date.* ("Up-to-date" and a few other compounds are always hyphenated, as the dictionary shows.)

As the first two of the preceding examples illustrate, when the second of the modifying words ("label" and "foot") is singular in number, the hyphen is always used. When the second word is plural ("labels" and "feet")—as shown in the second two examples—the hyphen is omitted (unless a compound like "up-to-date" is used).

The hyphen is also omitted in a compound adjective that begins with an adverb ending in -*ly*, as in "fully developed report."

You may need nothing more than common sense to know whether to use a hyphen or not. Note the difference between the two following statements in which the same compound adjective appears.

hyphen needed: Specifications call for $\frac{3}{4}$-inch *high-pressure* copper tubing.
hyphen not needed: Specifications call for *high pressure* copper tubing.

In the first sentence, the hyphen between "high" and "pressure" is needed so that the reader will not try to group "$\frac{3}{4}$-inch high" together. In the second sentence, he would have no problem because it would not be logical to group "pressure" and "copper" together.

Three other points about hyphenating and not hyphenating should be made:

1. You should use a hyphen to divide parts of an individual word if another meaning is possible *without* the hyphen.

   Example: Compare *re-sort* (to sort again) and *resort* (to have recourse).

2. You should use a hyphen to divide a word at the end of a line.

   Example: *mixture*

   Note that the division is always made at the end of a syllable (i.e., a sound).

3. You should not hyphenate chemical compounds used as compound adjectives.

   Example: *ammonium citrate* solution

(SP 4) Have Capitals Been Used (Not Used) Correctly? (cap)

When to Capitalize

You should always capitalize:

1. Proper names (*Consolidated Zinc Company*); words used in place of proper names (*Mother*); words used to designate a particular part of the country or world (*the South, the Far North*); names of countries (*Canada*); and names of languages (*French*).

2. Names for days of the week (*Monday*), months of the year (*October*), and holidays (*Christmas*).

3. Important words in titles of books, articles, etc. ("Increasing Production at the Johnson Corporation," "Silicones in Cosmetics"). Generally, articles, prepositions, and conjunctions of *less than five letters* are not capitalized.

4. Names of school subjects other than languages when a specific course is designated (*Physics 224, Geology 173*). (The name of a language is always capitalized, whether or not a specific course is designated.) Names and abbreviations of degrees (*Bachelor of Science, B.S.*).

5. The first word of a sentence, whether you are writing your own composition or are quoting the writing of someone else. ("Maintenance costs are increasing." "The manager's comment was, 'Maintenance costs are increasing.'")

6. People's titles when they are used as titles and not merely as describers ("Mr. John Talbott, Manager, Engineering, American Petroleum Corporation"; "Dr. Ralph Hastings, Professor of Chemistry, Eastern State University." *But* "Mr. John Talbott, a manager at American Petroleum Corporation"; "Dr. Ralph Hastings, a professor at Eastern State University").

When Not to Capitalize

On the other hand, you should not capitalize:

1. Words for *directions to* geographic areas. ("On my next trip east, I'll stop to see Mr. Winston.")

2. Names of seasons unless they are used as proper names ("spring" but "the Spring Festival exhibition").

3. Names of schools, vocations, etc., when a specific school, job position, etc., is not designated ("while I was in high school"; "since he is in engineering work").

4. Words used in place of proper names when they are qualified by a limiting adjective ("my mother," "this college"), or when a specific reference is not made ("a course in highway design").

EXPLANATIONS OF GRAMMATICAL TERMS (GT)

The following alphabetical list of grammatical terms, with definitions and illustrations, includes most of the common terms used in this book and elsewhere.

You will notice that for some terms reference is made to specific pages in your text. You should find enough information on these pages to understand the term fully in each case. If you wish more help, consult a basic English grammar text such as *Harbrace College Handbook.*[4]

Abstract noun See *Noun.*

Active voice See pp. 50–52.

Adjective An adjective is a word used to modify (change the essential meaning of) a noun. Adjectives are classified as *descriptive* or *limiting.*

---

[4] John C. Hodges and Mary E. Whitten, *Harbrace College Handbook*, 7th ed., New York, Harcourt Brace Jovanovich, 1972.

Descriptive adjectives express the nature or quality of nouns.

Examples: *clear* report, *tall* smokestack, *short* experiment, *brown* precipitate.

Limiting adjectives point out, restrict, or indicate the number or quantity of the nouns they modify.

Examples: *articles: a* memo, *an* obstruction, *the* generator.

demonstrative adjectives: this proposal, *that* method, *these* data, *those* circuits.

interrogative adjectives: whose letter? *what* building? *which* test tube?

numerical adjectives: two tests, *fifth* progress report.

possessive adjectives: my experiment, *your* research, *his* orders, *our* process, *their* decision.

Adjective clause An adjective clause is a subordinate clause used as an adjective.

Example: We accepted the proposal *that contained the least expensive solution.* (This clause functions as a demonstrative adjective. That is, the sentence could be abbreviated to "We accepted *that* proposal," since "that" points out "the proposal *that contained the least expensive solution.* ")

Example: The only test *that was successful* was the first. (This clause functions as a relative adjective; that is, it relates—refers—to the antecedent, or the words that precede, "only test.")

Adverb An adverb is a word used to modify (change the essential meaning of) a verb, an adjective, another adverb, or a main clause. Adverbs show *degree, manner, place,* or *time.*

Adverbs of degree indicate how much (His word was *much* better.).

Adverbs of manner tell in what way (He answered *quickly.*).

Adverbs of place tell where (We were sent *there.*).

Adverbs of time indicate when (The experiment is beginning *now.*).

Adverb clause An adverb clause is a subordinate clause used as an adverb.

Example: The experiment is beginning *while we are standing here.* (This clause functions as an adverb of time—telling when.)

Example: We were sent *where workers most needed our help.* (This clause functions as an adverb of place—telling where.)

Adverb clauses also indicate cause, comparison, concession, condition, manner, purpose and result:

The test failed *because the vacuum pump malfunctioned.* (cause)

His work was much better *than it had been.* (comparison)

We would buy the meter *even if it were more expensive.* (concession)

We will complete the experiment *if we have enough time.* (condition)

We began the procedure *as we had been instructed.* (manner)

The company is planning to add an extra shift *so that it can complete the work on time.* (purpose)

The company added an extra shift, *so that it was able to complete the work on time.* (result)

Agreement Agreement is the matching in gender, number, or person of words in a sentence.

Gender Gender tells the reader whether the noun or pronoun is masculine, feminine, or neuter (neither masculine nor feminine).

Masculine-gender agreement Example: John Hall sent *his* memo to the plant manager. (Since "John Hall" is masculine, "his" is the correct possessive pronoun.)

Feminine-gender agreement Example: Mrs. Carr resigned *her* job. (Since "Mrs. Carr" is feminine, "her" is the correct possessive pronoun.)

Neuter-gender agreement Example: The company presented *its* conclusions clearly. (Since "company" is neither masculine nor feminine—and therefore called "neuter"—"its" is the correct possessive pronoun.)

Number Number is the form of a noun or pronoun that tells the reader whether the noun or pronoun represents one or more than one person, place, or thing. Number is classified as singular or plural.

Singular-number agreement Example: Each is doing *his* own work. (Since "each" is singular—one person—"his" must be singular also.)

Plural-number agreement Example: They are doing *their* own work. (Since "they" is plural—more than one person—"their" must be plural also.)

Person Person is the form of a noun or pronoun that tells the reader whether the noun or pronoun is the speaker, the person spoken to, or the person spoken of. The speaker is called first person and is represented by *I* (singular) or *we* (plural). The person spoken to is called second person and is represented by *you* (singular) or *you* (plural). The person spoken of is called third person and is represented by *he, she,* or *it* (singular) or *they* (plural).

First-person agreement Example: I am doing *my* work. (The possessive pronoun "my" agrees with the singular first person "I.")

Second-person agreement Example: You did *your* best. (The possessive pronoun "your," which can be either singular or plural, agrees with the singular, or plural, second-person "you.")

Antecedent The antecedent is a word to which a pronoun refers.

Example: The *procedure that* was followed was quite simple. ("Procedure" is the antecedent of the relative pronoun "that.")

Example: After the *companies* merged, *they* began to consolidate the two operations. ("Companies" is the antecedent of the personal pronoun "they.")

Antecedents can also be phrases and clauses.

Appositive An appositive is one or more words that are placed next to another word and that add to or explain the first word. Both the word and the appositive have the same grammatical form or function.

Example: Mr. Tyson, *the plant manager,* requested that a study be carried out. (The appositive is "the plant manager"; it means the same as, and explains further, "Mr. Tyson.")

Article An article is a word, functioning as an adjective, that designates a noun or pronoun definitely or indefinitely. For a definite designation, "the" is used. For an indefinite designation, "a" or "an" (before a vowel) is used.

Examples: *The* report is on the desk. (A definite report is referred to.)

A report is being prepared. (Some report is referred to.)

Auxiliary verb An auxiliary verb is one that aids in expressing the aspect, mood, tense, or voice of another verb. It is therefore often called a "helping" verb.

Examples: He *will* leave soon.

I *ought* to go.

We *do* think the report is a little vague.

The most frequently used auxiliary verbs are *be, can, do, have, may, must, ought, should, will,* and *would.*

Case Case is the form of a noun or pronoun that tells the reader its function or relationship to other words in the sentence. Case is obvious in the use of most pronouns: simple subject (*I* went), compound subject (*He* and *I* worked together); simple object (John called *me*), compound object (John called *him* and *me*); simple possessive (I read *his* report), compound possessive (I read *his* and *your* report); however, the form of *you* and *it* is the same for subject or object.

Subjective (or nominative) case indicates a doer of action or something that functions to achieve some result or condition. (*Williams* did the experiment. The *report* is available. The *work* is complete.)

Objective case indicates a receiver of action. Direct objects receive action directly (He brought the *report.*). Indirect objects receive action indirectly (He brought the report to *me.*). Objects of prepositions qualify (modify, or change the meaning of) nouns, pronouns, verbs, adjectives, and adverbs (He went into the *office*). Possessive, or genitive, case shows ownership (*John's* experiment; *my* notebook) or—in a few uses where the apostrophe is also required—relationship (this *week's* project; a *day's* work—meaning "the project *of* this *week*"; "the work *of* a *day*).

Clause See pp. 43-47 for a complete explanation of clauses as they are used in the sentence. See also *Adjective clause, Adverb clause,* and *Relative clause* in this list of explanations of grammatical terms.

Collective noun See *Noun.*

Colloquialism A colloquialism is a word or expression that is used in informal speech and writing. See pp. 58-59 for several acceptable examples.

Common noun See *Noun.*

Complex sentence See pp. 46-47.

Compound-complex sentence See pp. 48-49.

Concrete noun See *Noun.*

Conjunction See p. 45 (coordinating conjunctions), p. 46 (subordinating conjunctions), and p. 47 (distinctions in use).

Conjunctive adverb A conjunctive adverb joins or relates main clauses. See pp. 47-48.

Demonstrative adjective See *Adjective.*

Demonstrative pronoun See *Pronoun.*

Dependent (Subordinate) clause See pp. 43 and 46.

Descriptive adjective See *Adjective.*

Direct address Direct address is a noun or pronoun, set off by commas, used to indicate that speech or writing is addressed to a particular person.

> Example: *Mr. Webb*, here is the letter you asked to see.

Direct object See *Object.*

Ellipsis Ellipsis is the omission of a word or phrase that leaves a sentence grammatically incomplete but nevertheless readily understandable.

> Example: The Red Valley plant produces more than the Allingham (plant produces). (The words "plant produces" are understood; the precise reader may even mentally fill in the sentence with them.)

Expletive An expletive is the idiomatic use of *it* or *there* to fill out a sentence.

> Example: *There* are 10 separate stages in this process.

Gerund See *Verbal.*

Gerund phrase See "Gerund" under *Verbal*, and then *Phrase.*

Idiom See pp. 53-54.

Indefinite pronoun See *Pronoun*

Independent clause (Main clause) See pp. 43, 44, 45-46, and 48.

Indirect object See *Object.*

Infinitive See *Verbal.*

Infinitive phrase See "Infinitive" under *Verbal*, and then *Phrase.*

Interrogative adjective See *Adjective.*

Interrogative pronoun See *Pronoun.*

Intransitive verb See *Verb.*

Limiting adjective See *Adjective.*

Linking verb A linking verb is a verb that expresses a condition or change without acting on an object. The most common linking verbs are *appear, be, become, feel, look, seem, smell, sound,* and *taste.*

> Examples: The process *is* inexpensive.
>   The work load *became* lighter.

Main clause See pp. 43, 44, 45-46, and 48.

Modifier A modifier is a word, phrase, or clause that limits or qualifies the meaning of another word, phrase, or clause. Modifiers are (or function as) adjectives and adverbs.

Modify To modify is to limit or qualify (change) the meaning of a word, phrase, or clause. See *Adjective, Adjective Clause, Adverb,* and *Adverb clause* for examples.

Mood See pp. 49-50.

Nominative (Subjective) See *Case.*

Nonrestrictive modifier See pp. 316-319.

Noun A noun is a word that names a person, place, thing, action, or quality. Nouns are identified as abstract or concrete, common or proper, and collective.

> An *abstract* noun names an idea, a quality, or anything else that has no material substance. (The *efficiency* of the worker was apparent.)

A *concrete* noun names anything that has material substance. (The *machine* had been in service without repairs for three months.)

A *common* noun names any class or any member of a class of things. (The *men* were all in the *laboratory*.)

A *proper* noun names a particular person, place, or thing. (The refinery is located in *Boston.*)

A *collective* noun is a singular form that names a group or collection. Its verb is singular if the group or collection acts or functions as one (The *staff is* at work in the laboratory); plural, if the group or collection acts or functions as separate individuals (The *staff are* planning *their* vacations).

Nouns function as:

Appositive (Mr. Ross, the *design engineer*, showed us how the machine worked.)

Indirect object of a verb (He explained the process to our *engineer.*)

Objective complement (We call the director of research a *genius.*)

Object of a gerund (Operating the *machine* is easy.)

Object of an infinitive (It is easy to operate the *machine.*)

Object of a participle (Conducting an *experiment*, he refused to be interrupted.)

Object of a preposition (The heart of the *system* is the adsorber.)

Object of a verb (He explained the *process* again.)

Person in direct address (*Mr. Ross*, please explain the process again.)

Predicate nominative (Mr. Wilson is the *president.*)

Subject of a verb (The *method* is well known.)

Noun clause A noun clause is a subordinate clause that functions, as a noun itself does, as an appositive, direct object, indirect object, object of a preposition, predicate nominative, or subject.

Appositive (The recommendation *that he drop the plan* annoyed him.)

Direct object (We conclude *that the process is efficient.*)

Indirect object (We gave *whoever asked* a complete set of directions.)

Object of a preposition (He accepted information from *whoever was willing to speak.*)

Predicate nominative (That was *what we recommended as a solution.*)

Subject (*What we recommended* was Plan B.)

Also possible structures, but used infrequently, are noun clauses functioning as an *objective complement* (We call "eager beaver" *whoever gets to work early)* and a *person in direct address* (*Whoever made that remark,* come forward.)

Number See "Number" under *Agreement.*

Object An object is a noun or pronoun, or a noun clause or noun phrase, that receives or is affected by a verb's action, or that follows a preposition. Objects are classified as direct object, indirect object, and object of a preposition.

Direct object of a verb Examples: He did write the *letter* (noun). He recommended *what had not been considered* (noun clause). The company approved *our taking time off* (noun phrase).

Indirect object of a verb   Examples: He sent *me* the manual (pronoun). He gave *whoever came in* a copy of the report (noun clause introduced by a pronoun). The company gave *those completing the course* a refund (noun phrase).

Object of a preposition   Examples: He brought instructions from *Al* (proper noun). Mr. Brown produced letters from *whatever companies the manager asked about* (noun clauses introduced by a pronoun). He quit without *finishing the job* (noun phrase introduced by a gerund).

Objective complement   An objective complement is one or more words added to the sentence to make the object complete. It gives more meaning to the object. For example: The manager named Mr. Frazier *our new supervisor.*

Participial phrase   See *Phrase.*

Participle   See *Verbal.*

Passive voice   See pp. 50-51.

Person   See "Person" under *Agreement.*

Personal pronoun   See *Pronoun.*

Phrase   A phrase is a group of two or more related words that does not contain a subject and a predicate. Phrases are classified as gerund, infinitive, participial, prepositional, and verb.

Gerund (noun) phrase   Examples: *Easing back on the throttle* is the first step in *shutting down the engine.* (The first five words are a gerund phrase functioning as the complete subject of the sentence. The last four words are a gerund phrase functioning as the object of the preposition "in.") See also *Verbal.*

Infinitive phrase   Examples: The first step is *to ease back on the throttle.* ("To ease back" is an infinitive phrase functioning as the predicate nominative; it means the same thing as "The first step" but is in the predicate, rather than the subject, of the sentence.) *To shut down* the engine, ease back on the throttle. ("To shut down" is an infinitive phrase functioning as an adverb telling why you—understood as the subject of "ease"—should ease back.) See also *Verbal.*

Participial phrase   Examples: "The man *easing back on the throttle* is the engineer. ("Easing back on the throttle" is a participial phrase functioning as an adjective to modify "man.") *Purchased in 1965,* the machine was still running well this year. ("Purchased in 1965" is a participial phrase functioning as an adjective to modify "machine.") All participial phrases function as adjectives. See also *Verbal.*

Prepositional phrase   Examples: We are working *with the latest equipment.* ("With the latest equipment" is a prepositional phrase functioning as an adverb modifying the verb "are working.") The machine *with the malfunctioning part* is *in the next room.* ("With the malfunctioning part" is a prepositional phrase functioning as an adjective modifying "machine." "In the next room" is a prepositional phrase functioning as an adverb modifying the verb "is.") See also *Preposition.*

Verb phrase Examples: We *have been working* with the latest equipment. ("Have been working" is a verb phrase functioning as the simple predicate of the sentence.)

For more information on phrases, see pp. 43 and 53-54.

Predicate The predicate is the verb in a sentence or clause along with its complements and modifiers. The predicate expresses what is said about the subject (states an action, indicates a change, represents a condition, etc.). Predicates are classified as simple and complete.

Simple predicates are verbs only (We *have been working* with the latest equipment. The process *is* short.).

Complete predicates are verbs plus complements and modifiers (We *have been working with the latest equipment.* The process *is short.*).

Preposition A preposition is a word that indicates the relation of a noun or pronoun to another word in the sentence.

Examples: The machine *with* the malfunctioning part is there. ("With" is a preposition that shows the relationship between the noun "machine" and the noun "part.") The plant closes *at* five o'clock. ("At" is a preposition that shows the relationship between the verb "closes" and the noun phrase "five o'clock.")

Prepositional phrase See *Phrase.*

Present tense See *Tense.*

Principal parts See *Verb.*

Pronoun A pronoun is a word that is used instead of a noun, noun phrase, or noun clause. Pronouns are classified as demonstrative, indefinite, intensive, interrogative, personal, reciprocal, reflexive, and relative.

Demonstrative pronouns point out their antecedents directly. *This* (singular), *these* (plural), *that* (singular), and *those* (plural) are the demonstrative pronouns.

Indefinite pronouns represent nouns generally or without indicating the number involved. Common indefinite pronouns are *all, any, anyone, anything, each, everyone, few, many, no one, none, one, some, someone,* and *something.*

Intensive pronouns supplement nouns and pronouns to give a statement emphasis. (Note that they are never used without a personal pronoun or a common or proper noun. Examples: I was there *myself.* The supervisor said so *himself.* Wilson did the work *himself.*) The intensive pronouns are *myself, yourself, himself, herself, itself, ourselves, yourselves* (plural), and *themselves.*

Interrogative pronouns introduce questions. *What, which,* and *who* are the interrogative pronouns. (Note that the sentence they introduce always ends with a question mark if a direct question is asked.)

Personal pronouns represent nouns that identify one or more persons.

In the subjective case, the personal pronouns are *I, you, he, she, it, we, you* (plural), and *they.*

In the objective case, the personal pronouns are *me, you, him, her, it, us, you* (plural), and *them.*

In the possessive case, the personal pronouns are *mine, yours, his, hers, its, ours, yours* (plural), and *theirs.*

Reciprocal pronouns are used to express mutual relationship. The most common reciprocal pronouns are *each other* and *one another.*

Reflexive pronouns are identical to the subject of the sentence. The reflexive pronouns are exactly the same as the intensive pronouns: *myself, yourself, himself, herself, itself, ourselves, yourselves* (plural), and *themselves.* One example illustrates the fact that they are used differently, however: He injured *himself* at work.

Relative pronouns introduce clauses that modify the antecedents. The relative pronouns are *that, which,* and *who.* Examples of their use are: This is the method *that* I was describing. The damaged machine, *which* had been removed from the shop, was still smoking. Moore is the person *who* gives the orders.

Proper noun   See *Noun.*

Reciprocal pronoun   See *Pronoun.*

Reflexive pronoun   See *Pronoun.*

Restrictive modifier   See pp. 316-19.

Sentence   See pp. 43 and 44-46.

Simple predicate   See *Predicate.*

Simple sentence   See pp. 44-46.

Subject   A subject is the word, phrase, or clause of a sentence about which the predicate states or asks something. *The tools* were dull. Subjects are classified as simple and complete. *Simple subjects* are single nouns or pronouns, or words functioning as nouns. Examples: The *worker* with the hat was promoted today. *He* is certain he is right. *Welding* is hard work.

Complete subjects are phrases and clauses—all the words about which the predicate states or asks something. Examples: *The worker with the hat* was promoted today. *He whom I was telling you about* is certain he is right.

Subjective complement   A subjective complement is one or more words added to the sentence to make the subject complete. It gives more meaning to the subject. Example: Mr. Frazier is an efficient *employee.*

Subjunctive mood   See pp. 49-50.

Subordinate clause   See pp. 43, 46, and 47-48.

Tense   Tense is the form of a verb that shows when the action, change, condition, etc., occurred. In English we have six active tenses: *present* (I *help,* I *am helping,* I *do help*), *past* (I *helped,* I *was helping,* I *did help*), *future* (I *will help,* I *will be helping*), *present perfect* (I *have helped,* I *have been helping*), *past perfect* (I *had helped,* I *had been helping*), and *future perfect* (I *will have helped,* I *will have been helping*). We also have passive forms for these tenses (I am helped, I am being helped, etc.)

See also p. 312 on the logic of the sequence of tenses.

Transitive See *Verb*.

Verb A verb is a word or group of words that express action, existence, or occurrence. (*Distill* is an "action" verb. *Be* is an "existence" verb. *Conclude* is an "occurrence" verb.)

As the dictionary shows for each verb listed, verbs have three principal parts: infinitive, past tense, and past participle. These parts are always shown in this order; examples are: (to) *give* (infinitive), *gave* (past), and *given* (past participle); (to) *take* (infinitive), *took* (past), and *taken* (past participle). Note that all verb forms are derived from these three parts.

Verbs are classified as transitive and intransitive.

Transitive verbs must have direct objects—expressed or understood—to complete their full meaning. An action is always carried over to a noun, a pronoun, or an element that functions as a noun.

Example: The foreman *assembled* the crew. (The action of the verb "assembled"—here used transitively—carries over to the direct object, "crew.")

Intransitive verbs need no direct objects to complete their meaning.

Example: The crew *assembled*. (The verb "assembled"—here used intransitively—takes no object; no action is carried over, nor could an action be carried over in this use of "assemble." An adverb *could* be added [The crew assembled *quickly*], but still no action would be carried over.)

Note that some verbs (such as "assembled") are both transitive and intransitive, but that when they are, they do not have the same meaning. As a transitive verb, for example, "assemble" means "to collect into one group." Thus, in the sentence "The foreman *assembled* the crew," the meaning is "caused the crew to collect into one group"; "crew" here is the direct object of "assembled." An action does carry over in *this* use.

Verb phrase See *Phrase*.

Verbal A verbal is a word that is derived from a verb and used as a noun, adjective, or adverb. Verbals act like verbs in that they can have subjects and objects. The three verbals are called "gerund," "participle," and "infinitive."

A *gerund* is a verb form that ends in *-ing* and functions as a noun.

Examples: *Experimenting* is a way of life for many specialists. (The gerund "experimenting" functions in this sentence as a noun, the subject of the sentence.)

He enjoyed *experimenting*. (The gerund "experimenting" in this sentence functions as the direct object of the transitive verb "enjoyed.")

He was tired of *experimenting*. (The gerund "experimenting" in this sentence functions as the object of the preposition "of.")

A *participle* is a verb form that ends in *-ing, -ed, -en* (rarely *-d, -n,* or *-t*) and functions as an adjective.

Examples: *Working* quickly, he finished the job on time. (The participle "Working" modifies the personal pronoun "he," which is the subject of the sentence.)

Overhauled after the last inspection, the machine is now running smoothly. (The participle "Overhauled" modifies the noun "machine," which is the subject of the sentence.)

Raising the flask, the technician signaled that the experiment was beginning. (The participle "Raising" modifies the noun "technician." Note that "flask" is the direct object of "raising.")

This is a *proven* theory. (The participle "proven" modifies the noun "theory.")

The manager was referring to *hidden* costs. (The participle "hidden" modifies the noun "costs.")

An *infinitive* is a verb form, usually preceded by "to," that functions as a noun, an adjective, or an adverb.

Examples: He wanted *to begin* the experiment. (The infinitive "to begin" introduces the verbal phrase "to begin the experiment." "To begin" functions as a noun, the direct object of the verb "wanted" and as a transitive verb taking the object "experiment.")

The foreman had a report *to write.* (The infinitive "to write" functions as an adjective modifying the noun "report.")

The process began *to look* difficult. (The infinitive "to look" functions as an adverb modifying the verb "began.")

An infinitive can also be passive.

Example: He wanted *to be helped.* The work *to be done* will take a great deal of time. We hurried *to be finished* in time.

Voice See pp. 50–51.

COMMON ERRORS IN GRAMMAR, DICTION, AND PUNCTUATION

On the following pages common errors in grammar, diction, and punctuation are identified, discussed, and corrected. Although the coverage is fairly broad, only the errors that occur most frequently are reviewed.

(SU) VIOLATIONS OF SENTENCE UNITY

(SU 1) Is the Sentence a Fragment (frag)

A sentence is called a fragment if it lacks a subject and a verb. The error is usually the result of careless, hurried writing. The writer leaves out subject or verb, or both; or he breaks a unified whole into parts that are separated by terminal punctuation (a period or semicolon used as a period).

fragment: Assistants in the laboratory not to accompany the project leader unless he can justify taking them with him *correction:* Assistants in the laboratory *are* not to accompany the project leader unless *he* can justify taking them with him.

correction: Assistants in the laboratory *may* accompany the employee *if* he can justify taking them with him.

Note the second correction above. It is included to show that a sentence may lack empathy as well as be grammatically wrong. Of course, many other revisions are possible.

fragment: The result was greater efficiency. Which was what our company needed.

correction: The result was greater *efficiency, which* was what our company needed.

fragment: The experiment was successful. Because great care had been taken throughout.

correction: The experiment was *successful because* great care had been taken throughout.

correction: The experiment *was successful because the technician had worked very carefully* throughout.

Again note the second correction. It is more effective than the first because an active structure ("the technician had worked") is always stronger than a passive one. See pages 50–51 for a discussion of active and passive voice.

(SU 2) Is the Sentence a Run-on? (run)

A sentence is called a run-on if two complete sentences are run together without a coordinating conjunction or without a terminal mark of punctuation (period, semicolon, question mark, exclamation point) between them.

run-on: He has taken several courses in programming thus he will be able to operate the computer without further training.

correction: He has taken several courses in *programming. Thus,* he will be able to operate the computer without further training.

correction: He has taken several courses in programming *and* thus will be able to operate the computer without further training.

correction: Because he has taken several courses in programming, he will be able to operate the computer without further training.

This correction may be even more effective. Observe that the more important idea, the last part of the sentence, is now the main clause. See pages 46–47 for a discussion of structuring sentences for ideas that are not equal in importance.

(SU 3) Does the Sentence Have a Comma Splice? (cs)

A sentence is called a comma splice if two main clauses (complete sentences) are joined by a comma without a coordinating conjunction (*and, but, for, nor, or*).

comma splice: The new machine was hard to operate, we were pleased to have it, though.

correction: The new machine was hard to *operate. We* were pleased to have it, though.

correction: The new machine was hard to *operate, but* we were pleased to have it.

The first correction is mechanically acceptable, you should note, but ineffective in comparison with the second.

(SC) VIOLATIONS OF SENTENCE COHERENCE

A sentence has coherence if (1) the reader finds that it is easy to read and understand, and (2) you as the writer logically fulfill the development indicated at the beginning of your sentence structure. The thought must be unambiguous (have only one meaning), and the relationship between words, phrases, and clauses must be clear.

(SC 1) Do the Subject and Verb Agree in Number? (agr)

The subject of your sentence controls the number of your verb. If the subject is singular, the verb must be singular. If the subject is plural, the verb must be plural.

disagreement: The nature of the variables were such that accurate work could not be done.

correction: The nature of the variables *was* such that accurate work could not be done.

disagreement: The supervisor and his crew was working overtime.

correction: The supervisor and his crew *were* working overtime.

The first sentence is incorrect because the writer failed to keep the singular subject (nature) in mind as he developed the sentence. The second sentence is wrong because the writer failed to remember that he began with a compound subject (two or more nouns joined by "and").

Note that two singular nouns in the subject usually—but not always—make a plural. If the nouns represent two people, *the* or some other limiting adjective should precede each noun. However, if the nouns identify one person who has *two* titles, no limiting adjective is used before the second noun.

correct: The president and *the* general manager *were* there.

(Two different people.)

correct: The president and general manager *was* there.

(One person having two titles.)

(SC 1a) Does a Phrase Intervene That Begins with "Together with," etc.?

A phrase that contains words understood to be subjects of the same verb, but that is set off by commas, does not influence the number of the verb.

disagreement: The engineer, along with the superintendent, have designed a different configuration.

correction: The engineer, along with the superintendent, *has* designed a different configuration.

(SC 1b) Are Correlatives Used with Singular Subjects?

If a set of correlatives ("either . . . or" or "neither . . . nor") is used with a singular subject, the verb must be singular.

disagreement: Neither the manager nor the supervisor were satisfied with the report.

correction: Neither the manager nor the supervisor *was* satisfied with the report.

If correlatives join a singular subject and a plural subject, the verb agrees in number with the subject closer to it.

disagreement: Either the men or the foreman were wrong.

correction: Either the men or the foreman *was* wrong.

(SC 1c) Does a Collective Noun Represent a Singular or Plural Subject?

A collective noun (crew, group, number, etc.) is considered singular in number when it acts as a unit, and plural when its constituents or components act individually.

disagreement: The number of employees sent to the new job were large.

correction: The number of employees sent to the new job *was* large.

disagreement: A number of employees was sent to the job.

correction: A number of employees *were* sent to the job.

(SC 2) Does Every Pronoun Agree in Number with Its Antecedent? (ant)

A pronoun stands in place of a noun. The number of the pronoun must therefore be the same as the number of its antecedent. If the number of the antecedent is singular, the number of the pronoun must be singular; if the number of the antecedent is plural, the number of the pronoun must be plural.

disagreement: He works for Zenith Furnace Company. They are the leader in the industry.

correction: He works for Zenith Furnace Company. *It is* the leader in the industry.

disagreement: Our laboratory has the newest facilities and equipment. A research organization needs this to be truly efficient.

correction: Our laboratory has the newest facilities and equipment. A research organization needs *these* to be truly efficient.

correction: Our laboratory has the newest facilities and equipment. A research organization needs *these advantages* to be truly efficient.

The second correction shows that a classifying word—like "advantages"—must often be used to make the thought immediately clear.

Pronouns whose antecedents are indefinite singular pronouns (*anyone, each, everybody, everyone, no one, somebody,* etc.) must also be singular in number.

disagreement: Everybody brought their lunch to work.

correction: Everybody brought *his* lunch to work.

disagreement: Each idea was analyzed in turn, but they were not set aside for further study.

correction: Each idea was analyzed in turn, but *it* was not set aside for further study.

correction: Each idea was analyzed in turn, but *none* was set aside for further study.

(SC 3) Is Every Pronoun Reference Clear? (ref)

A pronoun must not only agree in number with its antecedent, but also must show clearly to what or to whom it refers.

ambiguous: When the supervisor talked to the technician, he reviewed the problem in detail.

correction: When the supervisor talked to the technician, *the technician* reviewed the problem in detail.

correction: When the supervisor talked to *him*, the technician reviewed the problem in detail.

Note that the last correction does not repeat the antecedent (technician). Such repetition is a practice that is considered childish by many readers. You, too, can always find a way to write so that your thought is clear and your style is gracious.

ambiguous: When we found that quality control was substandard at our plant, we began comparing our procedures with those of similar companies. This is a major problem, we learned.

correction: When we found that quality control was substandard at our plant, we began comparing our procedures with those of similar companies. *Maintaining strict standards* is a major problem, we learned.

Writers who are lazy continually use *this*, alone, to refer to a previously stated idea. In the original sentence, *this* can mean that quality control is a major problem or that "comparing our procedures" is a major problem. When you are explicit you can help to make sure that your sentences are unambiguous.

confusing: In the review of the plant's research effort, it was found that it would be difficult to increase it because the research department was understaffed.

correction: In the review of the plant's research effort, *we* found that *we* would *have difficulty increasing* it because the research department was understaffed.

Of course, in the original only one *it* refers to the research effort; the first two *its* are expletives. However, in reading the original sentence for the first time, you will surely be confused.

vague: The process is then described, which leads to a discussion of the possible applications.

correction: The process is then described, *to help the reader understand* a discussion of the possible applications.

Which and the other relative pronouns (who, that) are used only to stand for nouns and other elements serving as nouns. Grammatically, the relative pronouns should not be used to stand for main clauses. "Which" above, for example, should not have been used to stand for the main clause, "The process is then described," in which the idea of the verb ("described") is all-important. Almost always, a revision such as the second sentence will be more effective as well as clearer and more correct. Here, for instance, the idea of helping appears to be much more meaningful than the idea of leading to.

(SC 4) Is There a Present Participle That Modifies a Main Clause? (trail)

Similar to the use of *which* to stand for a main clause is the use of a present participle to modify a main clause. A participle is a form of the verb used as an adjective; it can modify only a noun or another element serving as a noun.

trailing participle: We changed our manufacturing process, thus making sure that we would have a better finished product.

correction: We changed our manufacturing process. *Thus we made* sure that we would have a better finished product.

correction: By changing our manufacturing process *we made* sure that we would have a better finished product.

Analyzing the sentence, we find that *making* cannot modify the main clause ("We changed our manufacturing process.") However, the original sentence is also poor, we see, because the important idea should lie in the result rather than in the action that brings about the result. The idea of changing the manufacturing process is not important; what is, is the idea of making sure of a better finished product.

(SC 5) Is There a Dangler (an -ing or -ed Construction, or an Infinitive, That Modifies Nothing at All)? (dnlg)

Also found at the end of the sentence, but more often at the beginning, is a structure that does not modify anything. Such a structure is called dangling because it cannot be linked grammatically with a noun or another element functioning as a noun.

dangling participle: The distilling columns are on the right side when entering the laboratory.

correction: The distilling columns are on the right side, *you will find,* when *you enter* the laboratory.

correction: When *you enter* the laboratory, *you will find* the distilling columns on the right side.

dangling gerund: By cleaning up quickly, the shop will be ready for the next shift.

correction: By cleaning up quickly, *you will leave* the shop *so that it is* ready for the next shift.

correction: If you clean up quickly, *you will leave* the shop ready for the next shift.

Dangling constructions are often funny, as the preceding uncorrected sentences show. The first sentence seems to say the distilling columns are entering the laboratory; the second sentence seems to say that the shop itself should do the cleaning up. Both situations are ridiculous, of course. You can test for the incorrect use of a participle or gerund easily by placing it next to the noun that it modifies grammatically. To illustrate the test we shall rearrange the last original sentence.

The shop, by cleaning up quickly, will be ready for the next shift. . . . (impossible!)

Now we shall look again at the correction:

By cleaning up quickly, you will leave the shop so that it is ready for the next shift. . . . (logical)

The following sentences show that a past participle (verbal adjective ending in, for example, *-ed*) and an infinitive, respectively, can dangle, also:

dangling participle: When interviewed by your company's recruiter, he said that you were adding a new manufacturing division.

correction: When *I was* interviewed by your company's recruiter, he said that you were adding a new manufacturing division.

correction: When your company's recruiter *interviewed me*, he said that you were adding a new manufacturing division.

correction: When your company's recruiter *interviewed me, I learned* that you were adding a new manufacturing division.

dangling infinitive: To operate properly, the instructions enclosed must be followed.

correction: To operate *the machine* properly, the instructions enclosed *with it* must be followed.

correction: To operate *the machine* properly, *you must follow* the instructions enclosed *with it.*

(SC 6) Is Any Modifier in the Wrong Place? (mm)

Modifiers are labeled "misplaced" when they are not put where the reader can see their relationship to the words they logically go with.

misplaced modifier: Workers are frequently getting jobs approved by inspectors that are inferior.

correction: Workers are frequently getting *jobs that are inferior* approved by inspectors.

correction: Workers are frequently getting *inspectors to approve jobs* that are inferior.

Surely, no conscientious, qualified inspector would be pleased by the original sentence above!

Note, too, that single adverbs can create confusion when they are misplaced.

misplaced adverb: He only noted that the solution had been contaminated.
correction: Only he noted that the solution had been contaminated.

Always check carefully the position of such adverbs as *almost, also, even, just, merely,* and *only.*

(SC 7) Is There a Violation of Parallel Structure? (ll)

One of the most common errors made in writing is the use of grammatically different constructions to express similar ideas. This error is called violation of parallelism.

violation of parallel structure: We will be glad to send you details of our testing if you will write to the above address or by calling us at 717-294-5018.

correction: We will be glad to send you details of our testing if you will write to the above address or *call* us at 717-294-5018.

violation of parallel structure: In your study you will do well to review our custodial program under the following headings:
1. Electrical facilities
2. How good are our plumbing facilities?
3. Heating and ventilating
4. Our daily and weekly cleaning facilities must be thorough
5. Are rest-room and lounge facilities adequate?

correction: In your study you will do well to review our custodial program under the following headings:
1. Electrical facilities
2. Plumbing facilities
3. Heat and ventilating facilities
4. Daily and weekly cleaning facilities
5. Rest-room and lounge facilities

In the two sentences containing errors, the writers have begun with one construction and have then shifted to another. As the corrections show, like-items should be written in the same way (all noun elements, all verbals, etc.).

At times you have a choice of grammatical constructions when parallel structure is needed. For example, only questions could have been used in the last example. To set the questions up, the writer would have to revise the beginning to "In your study you will do well to seek answers to the following questions about our custodial program." The grammatical construction is always determined by the way you set up your presentation of like-items.

(SC 8) Are There Mixed Constructions? (mix)

A mixed construction is similar to the violation of parallel structure. Again the writer begins with one construction in mind, but he shifts to a construction that does not logically follow.

mixed: Because we approved, the changes in the blueprints enabled us to cut building costs.

correction: Because we approved the changes in the blueprints, *we were able* to cut building costs.

correction: Approving changes in the blueprints enabled us to cut building costs.

mixed: We are interested in learning more about your design, and what have you done to reduce manufacturing costs?

correction: We are interested in learning more about your design *and about your reduction of* manufacturing costs.

mixed: We are enclosing our bid, and advise us how soon after September 5 we can expect your order.

correction: We are enclosing our bid, and *we will appreciate your telling* us how soon after September 5 we can expect your order.

correction: We are enclosing our bid. *Please let us know* how soon after September 5 we can expect your order.

The corrections show that you can usually correct mixed constructions easily by checking the relationship of the larger parts (clauses, long phrases) to the basic thought of the sentence. The parts must be compatible with the basic thought, and the relationship must be logical if the sentence is to make sense.

(SC 9) Are the Tenses of the Verbs Compatible? (t)

The rule of the sequence of tenses is that your verbs and verbals must always be compatible. A little extra thought is usually all you need to make the tense of a verb or verbal in a qualifying clause or phrase logical.

illogical: We were glad to have received your report when it was due last week.
correction: We were glad *to receive* your report when it was due last week.
illogical: The purchasing agent promised that his company will review our bid.
correction: The purchasing agent promised that his company *would* review our bid.

In the first sentence the present perfect infinitive ("to have received") is not logical because it indicates a time prior to that of the action of the main verb ("were glad"). What the writer wants to show is that being glad and receiving the report when it was due occurred at the same time. To do this he must use the present infinitive ("to have").

In the second sentence the past tense ("would") must be used because the clause "that . . . bid" is an indirect statement of what was said in the past. In the actual conversation, the speaker said, "We *will* review, etc." But because the statement is indirect, the verb in the subordinate clause must show that the conversation took place at an earlier time.

(SC 10) Is Use of the Subjunctive Indicated? (subj)

The subjunctive mood is required after verbs of saying, demanding, recommending, suggesting, etc., when the tense of the verb following is *present.* Because the report writer uses the verbs *recommend* and *suggest* frequently, he should become familiar with this rule.

The subjunctive is also required when conditions are contrary to fact and when *doubt, wish, regret,* and similar verbs are used as verbs in the main clauses of complex sentences.

need subjunctive: It is suggested that the report is approved.
correction: It is suggested that the report *be* approved.
need subjunctive: If there was time, we would demonstrate the procedure now.
correction: If there *were* time, we would demonstrate the procedure now.

(SC 11) Are Too Many Subjunctives Used? (xSubj)

Some writers overuse the subjunctive. Frequently they insert an extra *would* in a sentence that expresses a condition.

extra subjunctive: We would produce more if we would overhaul our equipment.
correction: We would produce more if we *overhauled* our equipment.

The second subjunctive in the first sentence ("would overhaul") is incorrect because the condition, "if we overhauled our equipment," must be expressed as a definite action in order for the sentence to make sense.

To avoid using the extra subjunctive, a writer should try changing every clause that he begins with "if we would" to "were we to." He should ask himself, would anyone say, "would we be to overhaul," instead of, "were we to overhaul"? His answer would be, "no," unless he knew nothing about the correct idiom. Thus, he should be able to see that using the word *would* right after the conjunction *if* also makes the sentence sound illogical.

(D) ERRORS IN DICTION—WORD CHOICE AND USAGE

After you check your writing for grammatical errors, examine it to see if you have chosen words that are wrong for the context. Look especially for the following errors.

(D 1) Is Part of the Sentence Redundant? (r)

Redundancy—unnecessary repetition—is considered wrong because the reader naturally pauses when he sees an idea repeated; and if he ever pauses when he doesn't *have* to, the writer's communication of a message is weakened. Two types of redundancy are frequently found: the use of a verb that expresses an idea that has been stated already; and the use of a word or phrase that is basically the same in meaning as a word or phrase that has been presented.

redundant: Generally, the reduction in cylinder-liner wear fell in the range of 20 to 30%.
correction: Generally, the reduction in cylinder-liner *was from* 20 to 30%.
correction: Generally, the reduction in cylinder-liner wear *ranged from* 20 to 30%.
redundant: The meeting was attended by such leaders as the Director of Engineering, the Director of Research, the Director of Personnel, etc.
correction: The meeting was attended by such leaders as the Director of Engineering, the Director of Research, *and* the Director of Personnel.
correction: The meeting was attended by the Director of Engineering, the Director of Research, the Director of Personnel, etc.

redundant: The worker's statement was stated thoughtlessly.
correction: The worker's statement was *made* thoughtlessly.
correction: The worker *made the* statement thoughtlessly.

In the first sentence the writer is illogical in repeating the idea of reduction in the verb *fell.* In the second sentence the writer is illogical in using *etc.* when the idea of "and so forth" has already been presented by "such . . . as." In the third sentence the writer is illogical in repeating the idea of statement in the verb *stated*; repetition of this kind makes writing nonsense.

(D 2) Are Any Words Used Incorrectly for Other Words? (dict)

A number of writers often confuse words that are similar in sound or meaning. Examples are the use of *effect* for *affect, set* for *sit, lie* for *lay, imply* for *infer,* and vice versa.

wrong word: Oversupply effects prices.
correction: Oversupply *affects* prices.
wrong word: He lay the report on my desk.
correction: He *laid* the report on my desk.
wrong word: The missing screw was setting on the table.
correction: The missing screw was *sitting* on the table.

If you are not sure of the distinction between similar words, you should look them up in your dictionary. There, for example, you will find that the verb *effect* means to bring about. Thus, you will realize that a sentence like "Oversupply effects prices" is absurd. It is *affect* that means to influence or to cause to change. (You *can* say, "Oversupply *effects lower prices*," of course, but the qualifier *lower* is not included in the original sentence.)

Also in the dictionary, you will see that *lay* in the past tense comes from the verb *lie*, and that *lie* cannot take a direct object. (*Lie* is an intransitive verb, not a transitive one.) The verb required in the second original sentence is *laid*, which comes from the transitive verb *lay*. And in the dictionary you will find that *set* is a transitive verb. Thus, it cannot here be used in place of the intransitive verb *sit.*

Many writers use phrases incorrectly also, in particular "due to," "based on," and "similar to." These writers tend to use these phrases as adverbial modifiers, instead of a true adverbial element.

ungrammatical: Due to its formula, Ironcoat has several major limitations.
correction: Because of its formula, Ironcoat has several major limitations.
ungrammatical: Based on these data, we can predict that production this year will reach a new high.
correction: On the basis of these data, we can predict that production this year will reach a new high.

"Due to" and "based on" are both adjective phrases. Therefore they should be used only as modifiers of nouns and elements that function as nouns. The following examples show how these phrases can be used correctly:

Ironcoat has several major *limitations that are due to* its formula.
The new *high* predicted in production this year *is based on* these data.

"Because of" and "on the basis of," on the other hand, are adverbial phrases. Thus, they should be used in the original sentences because they modify verbs ("has" and "can predict," respectively).

(D 3) Is There Evidence of an Attempt to Be Too Concise? (con)

Leaders and workers, both, have a tendency to be too concise in using language, especially in writing to or for their colleagues. However, to be good idiom, fact and interpretation must be expressed fully. By being too concise, the writer forces the reader to fill in the gaps himself; and, from your own experience, you know that a reader should never have to supply words to get the message. (See Chapter 4, pp. 57-58.)

too concise: At the present time we are unable to approve your requested action.
correction: At the present time we are unable to approve *the action you re-quested.*
correction: At the present time we are unable to approve your *request for* action.
too concise: Production-wise, the Burleigh plant is lower than the Lewiston plant.
correction: The Burleigh plant is *below* the Lewiston plant *in production.*
correction: Production at the Burleigh plant is lower than *at* the Lewiston plant.

Use of combinations of a noun and the suffix *-wise* is especially worthy of note because they have become so ridiculous. Just consider the combinations that could be made in referring to parts of a classroom: *deskwise, blackboardwise, chalkwise,* etc.!

In looking over your own writing, check to see if some of your expression of facts and interpretations should not be elaborated—at least a little. You need to elaborate, of course, whenever extra empathy or tact is required, or the idea is especially important and the reader must not misread or skip over it.

(P) ERRORS IN PUNCTUATION

The use of correct punctuation—or the omission of it, if punctuation is neither needed nor wanted—is essential to clear communication. Indeed, if you neglect this adjunct to writing, your reader will surely be confused and frustrated.

To be sure, all writers cannot recognize their errors in punctuation merely by using pause tests. However, many writers can be more sure that they are punctuating correctly, if they use pause tests and common sense at the same time. When you want your reader to pause briefly, but not so long that he will feel that he has read a complete thought, use the comma, dash, colon, semicolon (when it replaces the comma), parenthesis, or bracket. When you want him to pause longer—that is, when you have presented a complete thought—use the period, semicolon (when it is used as a period), question mark, or exclamation point. When you want to punctuate for reasons other than indicating that the reader should pause, use quotation marks, italics, the apostrophe,[5] or hyphen.

The common errors that are discussed on the following pages should help you to recognize structures in which you tend to punctuate incorrectly. To be confident that you are using pause tests correctly, consult your dictionary or an English handbook. In the final analysis, punctuation is a matter of logic, of course, as the rules in these sources will show.

(P 1) Is a Comma (Are Commas) Used But Not Needed? (𝄐)

Clauses, phrases, and individual words are identified as either restrictive or non-restrictive elements. If an element is restrictive, a comma is not used. If an element is nonrestrictive, one or more commas must be used. Whether or not you need a comma depends on your thought. If the element is not necessary to give the sentence meaning, you should set the element off by commas. If the element is needed to give the sentence meaning, you should not use commas.

unnecessary comma: We frequently checked the circuits, which had been shorting unexpectedly.

correction: We frequently checked the *circuits which* had been shorting unexpectedly.

A restrictive element is indicated here because it is unlikely that *all* the circuits had been shorting. What the writer apparently meant was that two or three circuits (of, say, six) had been causing trouble. Hence, he and his colleague(s) checked the two or three.

When in such a situation should a comma be used where it appears in the original? The answer is that it should be used whenever it is clear that certain circuits, somehow identified as being troublesome, are being discussed. For example, if the sentence read, "We frequently checked the *primary* circuits, which had been shorting unexpectedly," the comma should be used. Note the italicizing of "primary" to show that the meaning of the sentence has been changed. Usually, the inclusion of qualifying words means "use a comma."

A better way (than is shown in the original sentence above) to indicate a restrictive element is to use the relative pronoun *that* instead of *which.* Almost always, *that* is used to tell the reader that the qualifier (relative clause) must not

---

[5] The apostrophe and hyphen are covered on pages 290-92.

be separated from the main clause ("We frequently checked the circuits"). Thus, the original would be clear without the comma and without the qualifying word, "primary," if it read: "We frequently checked the circuits *that* had been shorting unexpectedly." Remember: "that" indicates a restrictive (no comma) thought; indeed, most authorities call the relative pronoun "that" a *restrictive which.* (To refer to a person, these authorities would say, you should use the relative pronoun "who" whether the element is restrictive or nonrestrictive.)

Phrases and individual words can be restrictive, also.

unwanted commas: The number of reports, on this subject, is limited.
correction: The number of *reports on this subject is* limited.
unwanted commas: Our fellow technician, Walker, was not present.
correction: Our fellow *technician Walker was* not present.

In each of these sentences, commas will confuse the reader. "On this subject" in the first sentence tells him what reports, exactly, are being talked about. "Walker" in the second sentence tells him what fellow technician was not present. The sentences, respectively, indicate that there are other reports than those in the outer office, and that there are more fellow technicians than Walker. It is for these reasons that no commas should be used.

(P 2) Is There a Single Comma Between Subject and Verb or Verb and Complement? (✗)

Writers who do not distinguish between restrictive and nonrestrictive elements use the comma without thought. They use it especially between subject and verb, and verb and complement (a word or phrase after the verb that gives the sentence full meaning). They do so probably because they remember seeing sentences in which a comma appeared in one of these two parts of the sentence. However, unless they use a comma to prevent misreading, they should never insert a single comma in these places.

comma between subject and verb: These reports, are kept in the file in the plant manager's office.
correction: These *reports are* kept in the file in the plant manager's office.
comma between verb and complement: Then the superintendent decided, to change the work schedule.
correction: Then the superintendent *decided to* change the work schedule.

You should always look upon the element consisting of subject and verb, and the element consisting of verb and complement, as a *restrictive* construction—one whose meaning changes if a single comma is placed between the two parts. The only time a single comma is correct in one of these elements is when the sentence may be misread. Here is an example:

The worker referred to first, worked here ten years ago.

Naturally, if subject and verb or verb and complement should be separated by two commas, to set off a nonrestrictive element, both marks must be retained. This sentence illustrates:

> *These reports, which* are submitted *monthly, are* kept in the file in the plant manager's office.

Note the demonstrative pronoun "These." Functioning correctly, it points out what reports, exactly, are being referred to.

(P 3) Is a Comma (Are Commas) Needed But Not Used? (/,)

As was noted in section P 1, if the element is essential to the meaning of the sentence, commas are not used; the element is restrictive and is not set off from other parts of the sentence. If the element is not essential to the meaning, the element is nonrestrictive and commas are used; the element could be left out and the essential meaning of the sentence would not change. Otherwise, the only justification for using a comma is to prevent misreading. (See section P 2.)

comma needed: We frequently checked the primary circuits which had been shorting unexpectedly.

correction: We frequently checked the primary *circuits, which* had been shorting unexpectedly.

comma needed: As you know the manager of engineering at Eastern Forge will be here next week.

correction: As you *know, the* manager of engineering at Eastern Forge will be here next week.

commas needed: He said however much of the work had already been done.

correction: He *said, however, much* of the work had already been done.

commas needed: Five hundred of the workers or 47% said they read the company magazine regularly.

correction: Five hundred of the *workers, or 47%, said* they read the company magazine regularly.

The original sentences are confusing if commas are not used to guide the reader. The first sentence does not mean that some of the primary circuits had been shorting and some of them had not, but that is what the sentence says. The second sentence does not mean "As (because) you know the manager of engineering at Eastern Forge . . . ," but that is what the sentence says. The third sentence does not introduce a concession, but it seems to be starting a thought like "He said however much of the work had already been done, a great deal still remained." The last sentence does not mean "Five hundred of the sample" *or* "47% of some other group"—but that is what the original statement says. The reader has to review each sentence to understand it clearly. Because he must reread, something of the effectiveness of the writer's statement is lost.

Another rule is that a comma should always separate coordinate adjectives (adjectives that are equal in rank).

comma needed: The company's large fast truck will be used to deliver the machines.
correction: The company's *large, fast* truck will be used to deliver the machines.

You can see whether adjectives are equal in rank by applying either of two tests:
  1. Insert the word *and* between the adjectives.
  2. Interchange the adjectives.
If you find that the adjectives and the noun they modify make good sense after the test, you should use the comma. For example, here the phrase conceivably can be written "large and fast truck" or "fast, large truck."

Of course, you should use a comma only when the tests prove that the adjectives are coordinate. In the first sentence below you can quickly see that new and stake are not equal in rank:

unnecessary comma: The company's new, stake truck will be used to deliver the machines.
correction: The company's *new stake* truck will be used to deliver the machines.

The writer cannot logically say either "new and stake truck" or "stake, new truck." *Stake* restricts the meaning of truck so much that the original structure cannot be changed.

Frequently a comma is needed when word order is changed from the normal—that is, subject-verb-object-modifier. (See p. 52.) If the modifier comes first in the sentence (as it often does in a complex sentence), a comma is usually placed between the modifier and the subject of the main clause.

normal word order: The project supervisor will be pleased when he reads the latest progress report.
changed order requiring comma: When he reads the latest progress report, the project supervisor will be pleased.

Naturally, a comma is required after the main clause of a sentence presented in normal word order if misreading is possible when the comma is left out. However, if the sentence is very short, the comma may be omitted. It is almost always omitted when the subject of the subordinate clause is also the subject of the main clause, and the subordinate clause is extremely short.

When he *stopped he* immediately started to check the results.

(P 4) Is the Comma Used Incorrectly for the Dash, Parenthesis, or Bracket? (x/,)

The dash or parenthesis should be used instead of the comma when a series is inserted as an appositive, or is included in an adjective phrase following the word modified. The substitution is logical. Although the appositive or adjective phrase is nonrestrictive, the reader may not readily see where the element begins and ends if commas are used.

dashes or parentheses needed: Weather instruments, thermometers, barometers, and hygrometers, were the items we shipped to Jordan Laboratories.

correction: Weather *instruments–thermometers,* barometers, and *hygrometers– were* the items we shipped to Jordan Laboratories.

correction: Weather *instruments (thermometers,* barometers, and *hygrometers) were* the items we shipped to Jordan Laboratories.

The dashes or parentheses here enable the reader to retain mentally what immediately precedes the phrase, when he reaches the part of the sentence that follows it.

brackets needed: In his report, the technician wrote: "The only possible interruption, sic, is that the distillate is contaminated."

correction: In his report, the technician wrote: "The only possible interruption [sic] is that the distillate is contaminated."

Brackets are used primarily to identify errors in spelling, grammar, etc.–and thus to indicate that a mental correction must be made for clear understanding. In the original sentence above, commas will do nothing but confuse the reader. They will not–as brackets will–indicate that *interruption* was written erroneously for "interpretation." Usually, when brackets are used in this way, the writer is telling his reader, "This is exactly what the original I am quoting said."

Brackets are also used to supply needed word(s) for the integration of quotations with your own writing. ("[The use of automation] in this industry is widespread.") Sometimes, to clarify, the writer will add to–rather than replace– the quoted author's words. (During this year [1970], the company's profits began to decline.")

(P 5) Is the Semicolon Used Incorrectly for the Comma or Colon? (x/;)

(P 5a) Is the Semicolon Used Incorrectly for the Comma?

You may know that the semicolon is occasionally used in place of the comma. However, it should be used correctly in this way; today, a writer should substitute the semicolon for the comma only in the following situations:

1. The predicate of a sentence contains a series of phrases and internal commas appear within one or more phrases.

    Example: The tests that are to be made should include all of the physical tests that are needed: tests for specific properties, such as for protection against rust; tests for chemical values, for example, of water, sulfur, iodine, and acid; . . . .

2. The sentence is compound and the semicolon is used (instead of the comma) along with a coordinating conjunction. This use is logical because the presence of one or several commas within the first clause would tend to make the comma separating the clauses less obviously a divider of the two parts of the sentence.

   Example: The course was instituted for older employees, some of whom had been with the laboratory for twenty years; but recently young employees have dominated the enrollment.

3. The sentence is compound and the semicolon is used (in place of the comma) along with a coordinating conjunction. The reason for this use is to give greater emphasis to the two separate thoughts.

   Example: Most of the bidders for the contract responded at once; but several who had been successful bidders in the past waited until after the deadline.

The substitution is always made so that the reader will be able to continue without slowing down or missing the sign for a longer pause than the comma affords. Other substitutions of the semicolon for the comma—illustrated in the original sentences below—will be either confusing or ungrammatical.

confusing: Because this solution is time consuming, expensive to implement, and unpopular with our employees; we should examine other solutions immediately.

less confusing and correct: Because this solution is time consuming, expensive to implement, and unpopular with our *employees, we* should examine other solutions immediately.

ungrammatical semicolon: The reaction to the proposal, however; was generally unfavorable. *Or,* the reaction to the proposal; however, was generally unfavorable.

correction: The reaction to the *proposal, however, was* generally unfavorable.

A comma after "employees" in the first original sentence above is less confusing than a semicolon. Also, it enables the reader to go on quickly to the more important part—the main clause ("we should . . . immediately"). The semicolon today is regarded as a mark somewhere between the comma and the period; however, it is more often used as a period, as a terminal mark of punctuation rather than as a slowing one.

Keeping this usage in mind, you will not—as many do—maintain an image of a sentence containing *however* or another conjunctive adverb that is always preceded or followed by a semicolon. The image is maintained because punctuation for the conjunctive adverb can be difficult to master. By mentally replacing the semicolon with a period, you will realize that by incorrectly using the semicolon in this way you create a fragment in one part of the sentence unless the clauses are both main clauses.

(P 5b) Is the Semicolon Used Incorrectly for the Colon?

If you visualize the semicolon as a terminal mark of punctuation, you will not be tempted to use the mark as a colon. The colon is a mark that anticipates. It is used (1) before a series in apposition (usually right after the word that the appositive equals); (2) before a word, phrase, or clause (in the predicate) that explains or crystallizes the idea stated before it; and (3) in miscellaneous ways to show that more is to follow (minutes or seconds of time, body of a letter, etc.).

correct use for a series in apposition: The shipment just received from the Stewart Company contained standard paper products: bags, paper towels, and napkins.

correct use for explanation: We have stopped ordering from the E. C. Paul Company: its product has been unsatisfactory in recent months.

correct use for letter salutation: Dear Mr. Hodges:

correct use in numerals to show time: The consulting engineer will be here at 9:30 A.M.

Remembering the correct examples, you should be able to find the errors in the following sentences:

semicolon used for colon: Following is a list of the instruments that we have sold to the Northern Laboratory; barometers, hygrometers, thermometers, hydrometers, and pyrometers.

correction: Following is a list of the instruments that we have sold to the Northern *Laboratory: barometers*, hygrometers, thermometers, hydrometers, and pyrometers.

semicolon used for colon: Dear Mr. Hastings;

correction: Dear Mr. *Hastings:*

To distinguish the marks generally, always remember that the colon (:) anticipates and the semicolon (;) terminates.

(P 6) Are Quotation Marks or Italics Used Incorrectly? (x/") or (ital)

Following are the rules for citing publications and parts of publications. Use italics (of course, you must underline when typing or handwriting) for the title of a book or pamphlet, the name of a boat, the title of a painting, and names of other *whole* presentations. Use quotation marks for a chapter title, the title of an article, and other labeled items (such as a legal excerpt) that are only part of a presentation.

Writers often misuse italics and quotation marks. Especially, they tend to use quotation marks for all titles—probably under the influence of newspaper practice (newspapers do not set italics in the text of news stories).

incorrect: Dr. Bronson's latest book, "Chemistry in the 70's," is being highly praised.

correction: Dr. Bronson's latest book, *Chemistry in the 70's,* is being highly praised.

incorrect: More than 150 requests were received for reprints of the article *Silicones in Cosmetics.*

correction: More than 150 requests were received for reprints of the article "Silicones in Cosmetics."

Also incorrect is the writing of titles with no punctuation at all to mark them as titles.

Italics are used as well to emphasize a word or phrase, and for foreign words that are not in common use. Quotation marks are also used to enclose dialogue and slang, and words used in a special way or with a special meaning. However, the revised examples, above, illustrate the most common use of each mark.

(P 7) Are Punctuation Marks Incorrectly Placed Next to Quotation Marks (∩∪)

Writers should also be familiar with the rules for positioning commas, periods, and other marks next to terminal quotation marks. Periods and commas *always go inside* end quotation marks. Question marks and exclamation points precede the end quotes if they are part of the quotation, and follow if they are not. All other marks *always go outside* except in an unusual case (such as an article title ending in a dash). Examples of incorrect placement and corrections follow:

incorrect placement of comma and period: Before I could explain, the supervisor said, "Poor work", and then he added, "Come into my office later".

correction: Before I could explain, the supervisor said, "Poor *work,*" *and* then he added, "Come into my office *later.*"

incorrect placement of question mark: As instructed, I have read the article "Is Scientific Report Writing Adequate"?

correction: As instructed, I have read the article "Is Scientific Report Writing Adequate?"

incorrect placement of semicolon: Last week Mr. Williams quoted at length from the article "New Markets in the Electronics Industry;" then he discussed increases that could be made in our company's present markets.

correction: Last week Mr. Williams quoted at length from the article "New Markets in the Electronics *Industry*"; *then* he discussed increases that could be made in our company's present markets.

EXERCISES

ABBREVIATIONS

1. Correct the following abbreviations.
   a. 22 deg N.
   b. Med. Dr.
   c. Messers
   d. Professor John Mills, Dr. of Ph.
   e. + or –
2. Correct the following abbreviations.
   a. ac circuit
   b. 2 in board
   c. bar
   d. vap temp
   e. gal/min

NUMBERS

3. Correct the following numbers.
   a. One hundred and sixty million dollars
   b. Your letter of November fifteenth
   c. On the job were fourteen men
   d. One qtr. in pipe
   e. 82 Sixth Ave.

SPELLING

4. Correct the spelling errors in the following words.
   a. Very truely yours
   b. strech
   c. useing
   d. acheivement
   e. fullfill
   f. similiar
   g. feasable
   h. ajustment
   i. questionaire
   j. appropiate
5. Correct the spelling errors in the following sentences. Make any changes that are needed to prevent confusion.
   a. Our companies error was it's first of that kind.
   b. Everyone of those proccesses was allways acceptible in the passed.
   c. It definately was a well enginered project.
   d. Pore the sollution into two quart containers untill you have draned the vessel of all 64 ounces

e. It is the Mettalurgists responsability to complete this phaze of his asignment personaly.

EXPLANATIONS OF GRAMMATICAL TERMS

6. Indicate whether the following statements are true or false. Correct the statements that are false. (All statements are all-true or all-false.)
   a. A phrase is a group of related words that contains a subject and a predicate.
   b. An adverb is a part of speech that is used to modify a noun.
   c. A gerund is a verb form that ends in -ed and functions as an adjective.
   d. Ellipsis is the idiomatic use of *it* or *there* to fill out a sentence.
   e. Intransitive verbs must have direct objects—expressed or understood—to complete their full meaning.

COMMON ERRORS IN GRAMMAR, DICTION, AND PUNCTUATION

7. Identify and correct the violations of sentence unity in the following sentences. Improve each sentence only after making the basic correction.
   a. The results of this experiment are to be included in the second progress report. The one that is due next week.
   b. We were able to complete the test long before the deadline. Because we had worked every evening to compile and interpret the results of our experimentation during that day.
   c. The supervisors met alone first then they had a meeting with all the workers in the plant.
   d. As a result of the new directive there was even more absenteeism in the work force. Which was what the supervisor had predicted.
   e. He spent the morning testing the apparatus he had built, in the afternoon he ran the chemicals through for the first time.
8. Identify and correct the violations of sentence coherence in the following sentences. Improve each sentence only after making the basic correction.
   a. The results show clearly that each of the machines are not producing at full capacity.
   b. This problem, along with the problem we had been working on, were given top priority.
   c. Either the foreman or the men was wrong.
   d. The department's work on that problem was completed yesterday, and they have already begun a new project.
   e. Very little power is required to run the new machine in the finishing department. It is estimated to be less than two kilowatts.
   f. The roof is made of block gypsum covered with tar paper. It is doubtful that it would be strong enough to support it during a very severe ice storm.
   g. The total number of journeymen and apprentices on this job were greater than we had estimated.

    h. The magnetron is operating at a magnetic field density of less than 25,000 gauss, which is causing a major problem.

    i. The first phase of the project was completed in five hours, proving that we could save a great deal of time by using the new method.

    j. Firing the ceramic bodies at $1500°$ C, the stresses were relieved.

    k. By being thick, the negative pressure of extreme altitude will be negligible.

    l. Having been installed too quickly, our difficulties with the new equipment began the first day of operation.

    m. The fuel system of the dynamometer was drained, and then we put in regular gasoline to replace it.

    n. More accurate is a venturi will probably be the better unit for this purpose.

    o. We would have liked to have tested another sample of the product.

    p. Our recommendation is that the department hires another specialist in this field.

    q. If we would cool the product more quickly, we would increase its strength.

9. Identify and correct the errors in diction in the following sentences. Improve each sentence only after making the basic correction.

    a. All methods were partially successful in that they succeeded in producing a diffused layer.

    b. The excessive wear of these parts is sure to effect the overall performance of the unit.

    c. The main drive in both magnetos failed due to bending.

    d. Based on these data, we worked out an effective solution to the problem of overheating.

    e. Weather-wise, we should have a good day for our field trip tomorrow.

10. Identify and correct the errors in punctuation in the following sentences. Improve each sentence only after making the basic correction.

    a. That part of the project, which has not yet been completed, is the tabulation of results.

    b. The gages used in the strain gage work, have become more sophisticated.

    c. As he pointed out the time for conducting these tests is still far off.

    d. The proposal was discussed with Mr. Baldwin who makes the final decision on all requests for new projects.

    e. If the above sequence is repeated enough electrons can be freed from the material so that one can measure them as current.

    f. Three different tile materials, asphalt, vinyl, and rubber, were used for the tests.

    g. He ended his report with the note, "We need these valves right of way. Please ship at once."

    h. The following are needed for the experiment; an erlenmeyer flask, six test tubes 1" X 6", and a bunsen burner.

   i. Secondly, all conductors, including semiconductors, offer some resistance to current flow, and thus, the problem of supplying power and dissipating heat becomes a vital one.

   j. Since the work was being done by hourly employees; however, any change had to be negotiated with the union.

   k. Dear Dr. Kendrick;

   l. We could see the vapor form when the temperature was increased; all of the plastic lines and lucite parts in the injector system were crystal clear.

  m. The latest book on this subject, "High-Field Magnetrons", is now available downtown.

  n. His first comment was, "We are not interested;" then he weakened and told me to ask again next month, when, he said, "business may be better".

INDEX

Abbreviation
  in general, 280-81
  in letters, 151, 156, 157
  of state or territory, in letter, 151-52
  of technical terms, list of, 280-87
Abstract
  in formal report, 217
  informative, 121-24
  topical, 121-22
Abstract noun. *See* Noun
Accuracy, 11-13
Active voice. *See* Voice
Address
  envelope, 151-52
  inside, 151
  return, 150
Addressee line, 153
Adjective, 293-94
Adjective clause, 294
Adjustment letter. *See* Reply to claim
  letter
Adverb, 294
Adverb clause, 294
Agreement
  definition of, 295
  in gender, 295
  in number, 295, 305-06
Analogy
  as a deductive process, 84
  illustrated, extended, 4-7
  used to explain unfamiliar, 29

Announcement letter, 169-70
Antecedent
  agreement with, 306-07
  clarity of, 307-08
  definition of, 295
Apostrophe, 290-91
Appendix. *See* Proposal; Report
Application letter, for job
  attention-getting start in, 266-68
  example of, 269-70
  follow-up to, 271-72
  middle paragraphs of, 268-69
  objectives of, 260
  request for interview in, 269
Applied research. *See* Research
Appositive, 295; *see also* Noun; Noun
  clause
Arabic numeral. *See* Numeral
Article
  adapting to readers of, 250-57
  benefits of writing, 247-48
  body of, 252
  clearing, with employer, 258
  editorial statement for, example of,
    250
  ending of, 252
  finding a subject for, 248-49
  identifying readers of, 249-50
  introduction to, 251-52
  title of, 251
  writing, with regularity, 357-58

Article, definite and indefinite, 296
Asking questions. *See* Research
Attachment note, 156-57
Attention line, 155-56
Authorization, for report, 216-17
Auxiliary verb, 296
Awkwardness, 54-55

Bar chart, 130-31
Basis of classification, 108
Begging the question, 87
Bibliography, 79, 81-82, 196, 220
Block layout for letter, 149, 158
Bracket, 320
Brainstorming, 91-93

Candor, 17-18
Capitalization, 292-93
Carbon copy, 157
Case, 296
Cause-to-effect order. *See* Order
Caution note in description, 116-17
Chart. *See* Bar chart; Flow chart; Line
    chart; Segmented bar chart
Citing, 77
Claim letter, 166-68
Classification, 107-09
Clause
    adjective, 294
    adverb, 294
    definition of, 43
    dependent, 43, 46, 48
    independent, 43, 44, 45-46, 48
    main, 43, 44, 45-46, 48
    subordinate, 43, 46, 48
Clear Index, 61-63
Coherence
    of paragraph, 34-37
    of sentence, 48, 49, 305-13
Collective noun, and subject-verb agree-
    ment, 306; *see also* Noun
Colon, 322
Comma
    in changed word order, 319

for dash, 319-20
in nonrestrictive construction, 318-19
for parenthesis, 319-20
to prevent misreading, 317
with quotation marks, 323
in restrictive construction, 316-17
between subject and verb, 317
between verb and complement, 317
Comma splice, 305
Common noun. *See* Noun
Communication
    broad definition of, 4
    simple theory of, 4
Comparison order. *See* Order
Complaint letter, 166-68
Complement. *See* Subjective comple-
    ment; Objective complement
Complete predicate. *See* Predicate
Completeness, 15
Complex sentence. *See* Sentence
Complimentary close, 154-55
Compound sentence. *See* Sentence
Compound-complex sentence. *See*
    Sentence
Concession, 19
Conciliation, 19
Conciseness
    lack of, 16-17
    overdone, 315
Conclusion
    definition of, 70
    integration of, with illustration,
        139-40
    in reports, 195, 196, 197
    summary of conclusion(s), 127
Concrete noun. *See* Noun
Conjunction. *See* Coordinating con-
    junction; Subordinating conjunction
Conjunctive adverb, 47-48, 296
Contrast order. *See* Order
Coordinate adjective, 319
Coordinating conjunction, 45-46
Coordination, 45-46, 47
Correction symbol, 279-80

Correlative, and subject-verb agreement, 306
Courtesy title, in letter, 151
Covering letter, for job application, 266-72
Covering memo, 217
Cross-classification, 33, 108
Curve chart, 129-30

Dangler, 51, 309
Dangling participle, 51, 309
Dash, 319-20
Data sheet. *See* Résumé
Date line, in letter, 150
Decoding, 4-6
Deduction, 83-84
Definition
  definition of, 98
  in description, 109, 113, 114-15
  extended, 104-07
  pitfalls in, 102-04
  qualification of, 100-02
  in sentence, 99-102
Demonstrative adjective. *See* Adjective
Demonstrative pronoun. *See* Pronoun
Dependent clause. *See* Subordinate clause
Description
  definition of, 98
  of device, 109-12
  for observer, 112-15
  for participant, 115-17, 191-94
  of process, 112-17
Descriptive adjective, 294
Details-to-generalization order. *See* Order
Device description. *See* Description
Diction, errors in, 313-15
Differentia, in definition, 99, 100, 101
Direct address, 297; *see also* Noun clause
Direct object. *See* Case; Noun clause; Object

Documentation
  abbreviation in, 79-81
  bibliography for, 81-82
  footnotes for, 79-81
  reference system for, 82-83

Effect-to-cause order. *See* Order
Ellipsis, 297
Empathy
  in letters, 19-20, 149, 153, 160-61, 164-65, 170-71
  in reports, 18-19, 180
Emphasis
  in paragraph, 34-37
  in sentence, 48, 49
Enclosure note, 156-57
Encoding, 4-6
Endnote, 80n
English, standard, 56
Evaluation
  definition of, 70
  of ideas, 93
  by standards, 70, 88-90
Experimenting. *See* Research
Explanation, in letters, 161, 168
Expletive, 297
Extended definition, 104-07

Fact, definition of, 69
Familiar-to-unfamiliar order. *See* Order, of development
Feedback, 6
Figure. *See* Illustration
Figures. *See* Numbers
Flow Chart, 132
Fog Index, 60-61
Footnote, 79-81
Formal proposal. *See* Proposal, formal
Formal report. *See* Report, formal
Forms
  overall, 26-28
  uses of, 28-29
Fragment, 303-04
Full-block layout for letter, 149, 162

Gender, agreement in, 295
Generalization-to-details order. *See* Order
Genitive case. *See* Case
Genus, in definition, 99, 100, 101
Gerund. *See* Verbal
Gerund phrase. *See* Phrase
Gobbledygook, 59
Grammatical terms, 293-303
Graphic aids, 128-40

Hasty generalization, 85
Heading
  in letters, 150-51
  in reports, 31-33
  as topic head, 31-33
Hedging, 86
Homophone, 289
Hyphen, 291-92
Hypothesis
  definition of, 70
  in proposal, 183
  use of, 83, 91, 140

Idea
  brainstorming for, 91-93
  definition of, 71
  examples of, 71
  in place of conclusion, 140
Idiom, 53-54, 315
Illustration
  accuracy in, 138
  bar chart as, 130-31
  curve chart as, 129-30
  distortion in, 138
  figure as, 130-38
  flow chart as, 132
  integration of, with text, 139-40
  label for, 138-39
  line chart as, 129-30
  line drawing as, 134, 136
  map as, 132
  number of, 138-39
  photograph as, 134, 136

pictograph as, 134
preparation of, 136, 138-39
segmented bar chart as, 131
simplicity of, 136
source note for, 139
table as, 129
title of, 139
values, identification of, in, 139
Illustration as a form of development, 105
Imagination, 71, 91-93, 140
Imperative mood, 49-50, 116
Inaccuracy, 11-13
Incompleteness, 15
Indefinite pronoun. *See* Pronoun
Indicative mood. *See* Mood
Indirect object. *See* Case; Noun; Noun clause; Object
Induction
  explanation of, 83-84
  illustrated, with facts, 69-71
Infinitive. *See* Verbal
Infinitive phrase. *See* Phrase
Informal proposal. *See* Proposal, informal
Informal report. *See* Report, informal
Informational report. *See* Report
Informative abstract. *See* Abstract
Inquiry letter, 157, 158-60
Inside address, 151
Instructions for report, 216-17
Intensive pronoun. *See* Pronoun
Interpretation
  conclusions as, 70, 197, 220
  definition of, 69
  discussion of results as, 197
  of illustration, 140
  in procedure memo, 194
  in report, 194-97
Interpretative report. *See* Report
Interrogative adjective. *See* Adjective
Interrogative pronoun. *See* Pronoun
Interview. *See* Research
Intransitive verb. *See* Verb

Introduction. *See* Proposal; Report
Introductory summary, 125-26, 217
Inverted order. *See* Order
Invitation to bid letter, 165-66
Italics, 323

Jargon, 57-58
Job application. *See* Application letter
Justification of a conclusion, 90-91, 183

Language
  colloquial, 58-59
  connotative, 55
  denotative, 55-56
  everyday, 56-58
  false economy in, 57-58
  gobbledygook in, 59
  lively, 51-52
  polysyllabic words, 56-57
  slang, 58-59
  standard, English, 56-59
  stereotypes in, in letters, 171-74
  see also Words
Letter
  abbreviation in, 151, 156, 157
  for announcement. *See* Announcement letter
  of application. *See* Application letter
  attachment note in, 156-57
  attention line in, 155-56
  body of, 153
  carbon copy of, 157
  for claim. *See* Claim letter
  close of, 154-55
  common uses of, 28
  of complaint. *See* Complaint letter
  empathy in. *See* Empathy
  enclosure note in, 156-57
  explanation in, 161, 168
  heading of, 150-51, 154
  initials in, 156
  of inquiry. *See* Inquiry letter

inside address in, 151
for invitation to bid. *See* Invitation to bid letter
layout of, 149-57
letterhead of, 150
list in, 153-54
margins for, 149-50
for notice. *See* Notice letter
for order. *See* Order letter
as overall form, 26-27
paragraph in, 37, 153
reply to. *See* Reply to letter
of request. *See* Request letter
salesmanship in, 160-61
salutation of, 153
signature of, 155
subject line of, 156
Limiting adjective. *See* Adjective
Line chart, 129-30
Line drawing, 134, 136
Linking verb, 297
Lists, 37, 127-28, 153-54
Lively language, 51-52
Logic. *See* Research

Mail questionnaire. *See* Research
Main clause. *See* Clause
Map, 132
Margin. *See* Letter
Memo
  of authorization, 216-17
  common uses of, 28
  of instructions, 216-17
  as overall form, 27-28
  of procedure, 191-94
  as proposal, 180-91
  of transmittal, 191, 217
Misplaced modifier, 53, 310
Misstatement of problem, 86-87
Mixed constructions, 311
Modifiers, ungrammatical, 53, 297
Modify, grammar, 297
Mood, 49-50, 116, 194, 312-13
Multiple-choice question, 74-75

Negative words, 170-71
Nominative case. *See* Case
Nonrestrictive modifier, 314-17
Normal order. *See* Order
Note taking, 76-79
Notice letter, 169-70
Noun, 297-98
Noun clause, 298
Number agreement. *See* Agreement
Numbers, 287-88
Numeral, Arabic, 287-88
Numerical adjective, 294

Object, 298-99
Object of gerund. *See* Noun
Object of infinitive. *See* Noun
Objective of preposition. *See* Case;
 Noun; Noun clause; Object
Objective case. *See* Case
Objective complement. *See* Noun;
 Noun clause
Observer of process, 112-15
Observing. *See* Research
Open-end question, 75
Order, of development
 cause-to-effect, 30, 107
 comparison, 29
 contrast, 29-30, 34-35, 107
 details-to-generalization, 30, 34, 107
 effect-to-cause, 30
 elimination, 105
 etymology, 105
 examples, 105
 familiar-to-unfamiliar, 29
 generalization-to-details, 30, 34, 107
 illustration, 105
 simple-to-complex, 29
 space, 29
 time, 29
Order, in sentence
 inverted, 52
 normal, 52
 periodic, 52
 of phrases, 53

punctuation for changed, 319
Order letter, 161-62, 164-65
Orientation in description, 115-16
Outline, 32-34
Overlapping, 33, 108
Overwriting. *See* Conciseness

Paragraph
 coherence in, 34-35
 emphasis in, 34-35
 length of, 36, 37, 153
 poor development of, 17, 35-37
 unity in, 34
Parallelism, 310-11
Paraphrase note, 77
Parenthesis, 319-20
Participial phrase. *See* Phrase
Participle
 dangling, 309-10
 definition of. *See* Verbal
 trailing, 308-09
Partition, 107-08
Passive voice. *See* Voice
Periodic order. *See* Order
Person, agreement in, 295
Personal pronoun. *See* Pronoun
Photograph, 134, 136
Phrase
 defined, 43, 299
 functions of, 53
 kinds of, 299-300
 order of placement, 53
 wrong use of, 53-54
Pictograph, 134
Plagiarism, 77
Positive words, 170-71
Possessive adjective. *See* Adjective
Possessive case. *See* Case
Predicate, 300
Preposition, 300
Prepositional phrase. *See* Phrase
Present tense. *See* Tense
Pretesting. *See* Research
Primary source. *See* Source

Principal parts. *See* Verb
Problem solving, 87-88
Problem statement, 86-87
Procedures, 191-94
Process description. *See* Description
Professional title, in letter, 151
Progress report. *See* Report
Pronoun
    antecedent, agreement with, 306-07
    definition of, 300
    kinds of, 300-301
    reference to, 307-08
Proper name, capitalization of, 292
Proper noun. *See* Noun
Proposal, formal
    appendixes, 241
    cover of, 237
    introduction in, 237-38
    list of requirements in, 239-40
    plan of attack in, 239
    schedule of accounting for, 240
    statement of expected results, 240
    statement or technical approach in,
        238-39
    summary in, 240-41
    table of contents in, 237
    title page of, 237
    transmittal of, 237
    see also Proposal, informal
Proposal, informal
    for change, 181-83
    distinguished from report, 214
    introduction of, 191
    purpose of, 180-81
    for research, 183-91
    scope of, 191, 206
    see also Proposal, formal
Punctuation, common errors in, 315-23
Pure research. *See* Research
Purpose, statement of. *See* Proposal;
    Report

Question begging, 87
Question, idea-spurring, 92

Questionable authority, 85
Questionnaire. *See* Research
Quotation marks
    correct use of, 302
    incorrect use of, 302-03
Quotation note, 76

Readability formulas, 59-64
Reader
    analogous to receiver, 4-7
    inaccurately identified, 12-13
    wants of, 7-8
Reciprocal pronoun. *See* Pronoun
Recommendations, 30, 31, 195, 196,
    197, 220
Redundancy, 313-14
References
    in documentation, 83-83
    in résumé, 265
Reflexive pronoun. *See* Pronoun
Relative adjective. *See* Adjective
Relative pronoun. *See* Pronoun
Reply to claim letter, 168-69
Reply to complaint letter, 168-69
Reply to inquiry letter, 160-61
Reply to request letter, 160-61
Report, formal
    abstract of, 217
    appendix of, 220
    authorization for, 216-17
    background statement in, 219
    conclusions in, 220
    cover of, 215
    covering letter for, 217
    definition of terms in, 219
    instructions for, 216-17
    introduction of, 218-19
    introductory summary of, 217
    methods of collecting data, state-
        ment of, in, 219
    purpose, statement of, in, 218
    recommendations in, 220
    report proper of, 219
    scope, statement of, in, 218-19

Report, formal (*Continued*)
  table of contents of, 216
  title of, 216
  title page of, 215-216
  transmittal for, 217
  see also Report, informal; and individual entries
Report, informal
  abbreviations, technical, in, 281-87
  abstract of, 121-24
  appendix of, 196
  basic structures for, 195
  classification in, 107-08
  common uses of, 28
  conclusions in, 127, 139-40, 195, 196, 197
  definition of, 8
  definition, sentence, in, 99-104
  description in, 108-17
  discussion of results in, 197
  distinguished from proposal, 214
  documentation for. *See* Documentation
  evaluation in, 88-90
  extended definition in, 105-06
  figure in. *See* Illustration
  illogic in, 84-87
  inaccuracy in, 11-13
  informational, 194-96
  interpretive, 195-97, 219
  introduction of, 195
  introductory summary of, 125-26
  justification in, 90-91
  list in, 127-28
  logic in, 83-84
  outline of, 32-34, 195
  as overall form, 27-28
  partition in, 107-08
  problem solving in, 87
  problem statement for, 86-87
  of progress, 205
  purpose, statement of, in, 13-15, 196, 206-08
  recommendations in, 30, 31, 195, 196, 197

  research for. *See* Research
  results of study in, 197
  review of present work in, 197
  review of previous work in, 197
  scope, statement of, in, 191, 206-08
  summary of, 126-27, 196
  table in. *See* Illustration
  terminal summary in, 126-27
  text summary in, 126
  see also Report, formal; and individual entries
Representative sampling. *See* Sampling
Request letter, 157-58
Research
  applied, 71
  asking questions in, 73-75; *see also* Inquiry letter
  definition of, 69
  experimenting in, 73
  illogic in, 84-87
  imagination, use of, in, 71, 91-93
  interpretation in, 69-71, 87-91, 140
  interviewing in, 73-75
  logic in, 83-84
  observing in, 73
  pretesting in, 75
  proposal for, 183-91
  pure, 71
  questionnaire, use of, in, 73-75
  sampling in, 75-76
  tabulating results of, 76
Restrictive modifier, 316-19
Résumé
  definition of, 260
  details of work experience in, 265
  heading in, 263
  information about schooling in, 263-64
  names of references in, 265
  objectives of, 260
  personal data in, 265
  plans for further training in, 264
  qualities of, 262
  statement of goals in, 263
  style of, 265

Run-on sentence, 304

Salesmanship in letters, 160-61
Sampling
  representative, 75-76
  weighted, 75-76
Scope. *See* Proposal; Report
Secondary source. *See* Source
Segmented bar chart, 131
Semiblock layout for letter, 149, 159
Semicolon
  and comma splice, 304
  confusing use of, 321, 323
  correct use of, 320-21
  ungrammatical use of, 321, 323
Sentence
  awkwardness in, 54-55
  complex, 46-47
  complex-compound, 48-49
  compound, 45, 47
  defined, 43
  mood of verb in, 49-50, 194
  order in, 52
  simple, 44-45
  voice of verb in, 50-52
Sentence definition. *See* Definition
Sentence head in outline, 32
Simple predicate. *See* Predicate
Simple sentence. *See* Sentence
Simple-to-complex order. *See* Order
Simplified layout for letter, 149, 163
Source
  primary, 73-76
  secondary, 71-72
Space order. *See* Order
Specifications, 88-89
Spelling
  test of, 288-89
  use of apostrophe in, 290-91
  use of capital in, 292-93
  use of hyphen in, 291-92
  words in list, 289-90
Standards
  for evaluation, 88-90
  importance of, 91

as specifications, 88-89
Statement of significance
  in device description, 109
  in process description, 113
Stereotypes in letter writing, 171-74
Subclassification, 108
Subject
  agreement with verb, 305-06
  of sentence, 301
Subjective case. *See* Case
Subjective complement, 301
Subjunctive mood. *See* Mood
Subordinate clause. *See* Clause
Subordinating conjunction, 46
Subordination, 46-47
Substantial majority, 86
Summary
  abstract as, 121-24
  of description, 111
  in formal report, 217
  in informal report, 195, 196
  introductory, 125-26
  omission of, 20-21
  suggestions for writing, 123-24
  terminal, 126-27
  text, 126
Summary note, 77
Sweeping generalization, 85

Table, 129
Tabulating. *See* Research
Tense
  definition of, 301
  sequence, rule of, 312
Term, in definition, 99, 100, 101
Terminal summary, 126-27, 196, 220
Text summary, 126
Time order. *See* Order
Title
  of article, 251
  of formal proposal, 237
  of formal report, 215-16
  of illustration, 139
Tone
  of job application letter, 268

Tone (*Continued*)
  of letter generally, 170-71
Topic head in outline, 32
Topic sentence
  development for, 34
  ensures coherence, 34
  ensures emphasis, 34
  ensures unity, 34, 36
  as generalization of fact, 195-96
  as interpretation, 195-96
Trailing participle, 308-09
Transition, 34-35
Transitive verb. *See* Verb
Transmittal, 191, 217
Two-way question, 74

Underlying principle in description
  of device, 110
  of process, 113
Unity
  of paragraph, 34-37
  of sentence, 48, 49, 303-05

Verb, 302
Verb phrase. *See* Phrase
Verbal, 302-03
Voice, 50-51

Warning note in description, 116-17
Weighted sampling. *See* Sampling
Wishful thinking, 85
Word order. *See* Order, in sentence
Wordiness, 16-17
Words
  confused often, list of, 56
  connotative, 55
  denotative, 55-56
  negative, 170-71
  order of, in sentence, 53
  polysyllabic, 56-57
  positive, 170-71
  ungrammatical use of, 314-15
  wrong, for context, 53-54
Writer, analogous to sender, 4-7

A 2
B 3
C 4
D 5
E 6
F 7
G 8
H 9
I 0
J 1